ESSENTIALS OF STRUCTURAL EQUATION MODELING

Dr. Mustafa Emre Civelek

Zea Books
Lincoln, Nebraska 2018

ISBN: 978-1-60962-129-2

doi: 10.13014/K2SJ1HR5

Zea Books are published by the University of Nebraska–Lincoln Libraries.

Electronic (pdf) editions published online at
https://digitalcommons.unl.edu/zeabook/

Print editions sold at
ttp://www.lulu.com/spotlight/unllib

UNL does not discriminate based upon any protected status.
Please go to unl.edu/nondiscrimination

Nebraska
UNIVERSITY OF
Lincoln

PREFACE

Structural Equation Modeling is a statistical method increasingly used in scientific studies in the field of Social Sciences. Nowadays, it is a preferred analysis method especially in doctoral dissertations and academic researches. However, since most of the universities do not include this method in the curriculum of undergraduate and graduate courses, students and scholars try to solve the problems they encounter by using various books and internet resources. The aim of this book is to guide the researcher who wants to use this method in a way that is free from math expressions and to teach the steps of a research using structured equality modeling practically. For the students who write thesis and the scholars who prepare academic articles, this book aims to systematically analyze the methodology of the scientific studies which is conducted by using structural equation modeling method in social sciences. This book is prepared in a simple language as possible so as to convey basic information. This book consists of two parts. In the first part, basic concepts of structural equation modeling are given. In the second part, examples of applications are given.

CONTENT

1. INTRODUCTION

There are various computer programs that are used when the structural equation modeling method is applied. The most common ones are LISREL (Linear Structural Relations), AMOS (Analysis of Moment Structures), MPlus, EQS (Equation Modeling Software) (Taşkın & Akat, 2010). The samples and illustrations in this book were made according to the AMOS program. The AMOS program is a visual program that is easier to use than other programs. This is why it is preferred in this book. There will also be plenty of videos related to AMOS on YouTube and similar sites. That will make the learning process easier for readers. There are different versions of the AMOS program. AMOS 22 version was used for the analysis examples in the book. All the narrations in the book have been tried to be supported by visuals. In this book, it is assumed that the readers have basic statistical information. The main aim is to provide a guide to the readers on structural equation modeling.

Although the structural equation modeling method is similar to linear regression analysis, it has many advantages. Some of the features that outperform the structural equation modeling are summarized below. These superior features distinguish structural equation modeling from other classical linear modeling approaches (Çelik & Yılmaz, 2013).

1. It reveals the relationship among hidden structures that are not directly measured.

2. Possible mistakes in the measurements of the observed variables are taken into consideration. The classic regression approach assumes no measurement error.

3. It is a very useful method to analyze highly complex multiple variable models and to reveal direct and indirect relationships between variables.

In this book, firstly the basic concepts related to structural equation modeling are discussed. Basic concepts such as measurement model versus structural model, latent versus observed variables, and endogeneous versus exogeneous variables are explained. Once the goodness of fit indices have been defined, all the processes of a research have been dealt with up to hypothesis testing, from determining the validity and reliability of the scale. The first chapter of the book consists of a description of the topics related to the structural equation modeling. In the second chapter, there are sample applications. Samples consist of summaries of actual research done on the field. This is intended to give readers a template that they can follow in their own research. It is recommended that the readers who will be using AMOS for the first time have started this book after looking at the tutorial videos on YouTube and similar sites. Also, to use AMOS, it is absolutely necessary to be able to use SPSS and to have basic statistical information. Assuming that the readers have this information, the subject explanations have been prepared.

I. CHAPTER

CONCEPTS AND METHODOLOGY

2. BASIC CONCEPTS OF STRUCTURAL EQUATION MODELING

2.1. Definitions and Features

Structural equation modeling is a statistical method increasingly used in scientific studies in the field of social sciences in recent years. The most important reason of the spread of this statistical technique is that the direct and indirect relationships among causal variables can be measured with a single model (Meydan & Şen, 2011). Structural equation modeling is a statistical method used to test the relationships between observed and latent variables. Observed variables are the measured variables in the data collection process and latent variables are the variables measured by connecting to the observed variables because they can not be directly measured. Structural Equation Modeling consists of two basic components as structural model and measurement model.

Another reason for the widespread adoption of this method is the ability of taking in to the account of the measurement errors and the relationships between errors in the observed variables. In this way measurement errors can be minimized. In traditional regression analysis, potential measurement errors are neglected. Another difference from the regression models of structural equality models is that they are based on the covariance matrix. For this reason, in some sources, it is named as covariance structure modeling or analysis of covariance structure (Bayram, 2013). On the other hand, , the correlation matrix is the basis of the regression. Covariance is a nonstandardized measure of the relationship between two variables, so it can take values between $-\infty$ and $+\infty$. Correlation, however, can take values between -1 and +1, since it is standardized (Gujarati, 1999). The covariance which is the basic statistic of the structural equation model can be shown for two observed and continuous variables as follows:

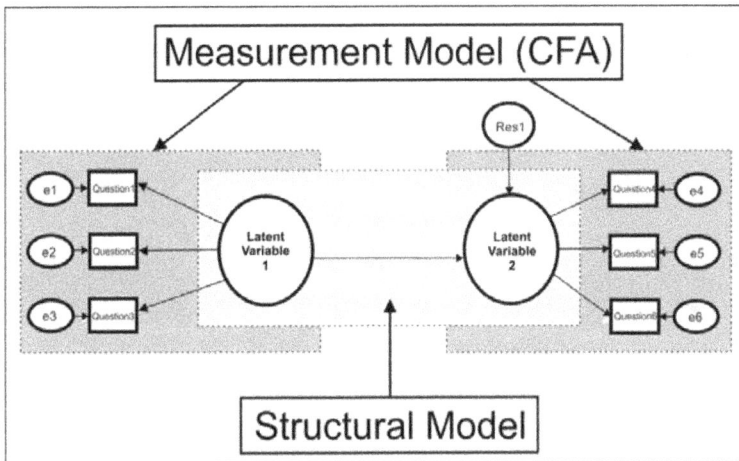

Figure 1. Demarcation between Measurement Model and Structural Model

Source: Byrne, B. M. (2010). Structural Equation Modeling with AMOS. New York: Routledge Taylor & Francis Group.

$$Cov_{xy} = r_{xy} \, SD_x \, SD_y$$

In the formula shown above, r_{xy} indicates the Pearson correlation coefficient, and also, SDx and SDy indicate the standard deviation of each variable (Kline, 2011). Structural equation modeling differs from some other multivariate statistical methods in that it is a confirmatory approach. In Table 1, it can be seen that the explanatory and confirmatory ones of the multivariate methods are (Hair, Hult, Ringle, & Sarstedt, 2017).

Tablo 1. Explanatory and Confirmatory Multivariate Methods

Explanatory	Confirmatory
• CulusterAnalysis	• Analysis of Variance
• Explanatory Factor Analysis	• Logistic Regression
	• Multiple Regression
• Multidimensional Scaling	• Confirmatory Factor Analysis
• Partial Least Squares Structural Equation Modeling (PLS-SEM)	• Covariance Based Structural Equation Modeling (CB-SEM)

Source: Hair, J., Hult, G., Ringle, C., & Sarstedt, M. (2017). A primer on partial least squares structural equation modeling PLS-SEM. Los Angeles: SAGE.

Most of the statistical methods other than structural equation modeling try to discover relationships through the data set.

However, structural equation modeling confirms the correspondence of the data of the relations in the theoretical model. For this reason, it can be said that structural equation modeling is more suitable for testing the hypothesis than other methods (Karagöz, 2016). Structural equation modeling consists of a system of linear equations. The key in the regression analysis is to determine how much of the change in the dependent variable is explained by the independent variable or variables. Although multiple regression analysis can only be applied to observed variables, the basic principles can be applied to structural equation modeling (Kline, 2011). Differently from the regression, structural equation modeling, as a new statistical analysis technique, allows to test research hypotheses in a single process by modeling complex relationships among many observed and latent variables. In traditional regression analysis, only direct effects can be detected. However, in the method of structural equation modeling, direct and indirect effects are put together.

In order to test the accuracy of the conceptual model, the most common method encountered in the literature on structural equation modeling is a two-stage method consisting of measurement model and structural model. In the first stage, the measurement model is tested; in the second stage the structural model is tested. The measurement model measures how well hidden variables are represented by the observed variables. It is mainly confirmatory factor analysis (CFA) and indicates the contruct validity of scales. Therefore, if the measurement model fit indices are low, it will not make sense to test the structural model (Dursun & Kocagöz, 2010). As seen in Figure 1, structural equation modeling is a compound of factor analysis and regression analysis. The measurement model and the structural model are interwoven. But the structural equation modeling is based on the confirmatory approach. It is based on the statistical confirmation of the theoretical model. For this reason, the measurement model is confirmatory factor analysis.

2.2. Latent and Observed Variables

The obseved variable (manifest variable) is the measured variable in the data collection process; latent variables are variables that are measured by connecting to observed variables because they can not be directly measured. Latent variables must be represented by more than one observed variable since they represent abstract concepts. It is recommended that the number of observed variables connected to a latent variable in structural equality models be at least three. Latent variables represent hypothetical constructs in a research model (Raykov & Marcoulides, 2006).

Observed variables in the structural equation modeling can be categorical, discrete or continuous variable, but latent variables must always be continuous variable. There are other statistical techniques that allow analysis with categorical latent variables, but in structural equation modeling only continuous variables can be analysed (Kline, 2011). Continuous variable is a variable that takes any random value from the set of real numbers. However discrete variable can only take value from the set of even number. Research questionnaires used to collect data in social sciences are generally prepared by using the Likert type ordinal scale. Values in this scale type take integer values ordered by importance. Therefore, the indicators used to describe a concept are discrete variables.

Latent variables are always continuous variables due to they are linked to more than one indicator.

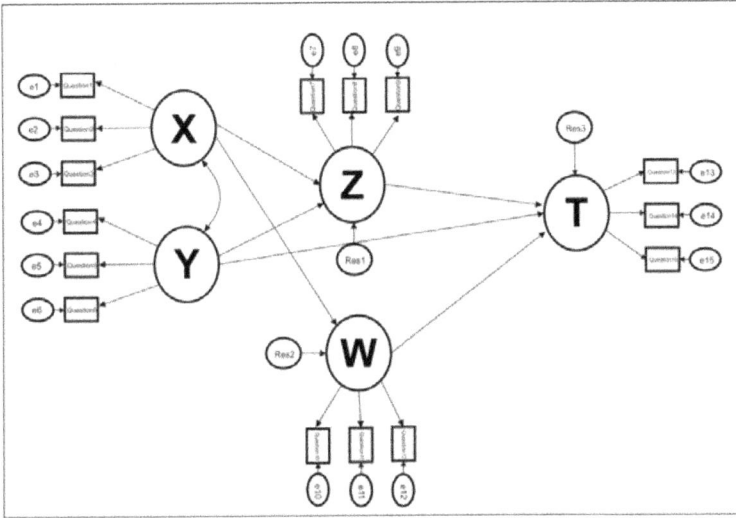

Figure 2. Endogeneous and Exogeneous Variables

As shown in Figure 1, the first latent variable (Latent Variable 1) is linked to the observed variables (Question 1, Question 2, and Question 3) which are composed of the questions answered in the research questionnaire. As seen in Figure 1, the observed variables and latent variables are connected to each other in a reflective way.

In the same way, the observed variables of the second dimension (Question 4, Question 5 ve Question 6) in the conceptual model of the research are connected to another latent variable (Latent Variable 2). The direction of the arrows connecting the observed variables and the hidden variables is important.

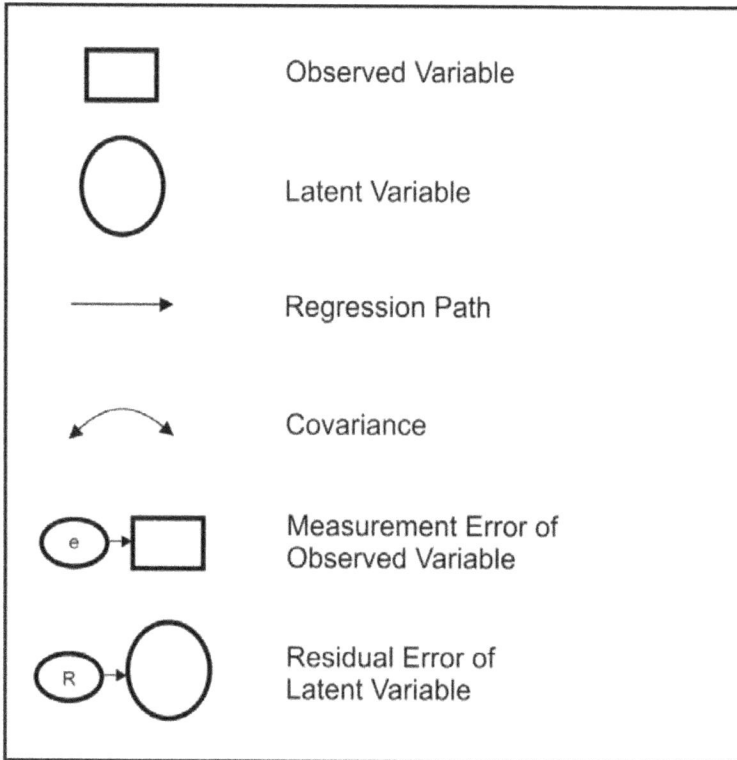

▭	Observed Variable
◯	Latent Variable
⟶	Regression Path
⌢	Covariance
e→▭	Measurement Error of Observed Variable
R→◯	Residual Error of Latent Variable

Figure 3. Symbols in Structural Equation Models

Again referring to the examples in Figure 1 and Figure 2, the quadrangular symbols indicate the observed variables, while the elliptical symbols represent the predicted latent variables (this book is based on the notation used in the AMOS program). Figure 3 shows the meanings of the most commonly used symbols in structural equation models.

2.3. Endogeneous and Exogeneous Variables

Variables in the structural equation modeling, except for the distinction between latent and observed, are dealt with in two groups: endogeneous and exogeneous. In structural equation modeling, the Endogeneous Exogeneous distinction is used as a more accurate distinction because a variable can assume the role of both the dependent variable and the independent variable at the same time. Endogeneous variables are dependent variables explained by other variables. In Figure 2, the variables Z, W and T are endogeneous variables. Exogeneous variables are independent variables that are not explained by any variables. In Figure 2, the variables X and Y are external variables. If there are more than one exogeneous variable, covariance between these variables is required. As shown in Figure 2, a bi-directional arrow is placed between the X and Y latent variables. It should not be forgotten to add an error term to the endogeneous variables. The terms Res1, Res2 and Res3, which appear in Figure 2, represent the residuals of each endogeneous variable. These residuals are also called as error terms of the structural model. The error terms in the measurement model are shown in Figure 2 by the notation 'e'. The error terms in both groups are usually marked with a separate notation in this way. Unlike regression models, structural equality models are based on the covariance matrix. But it mainly consists of the system of linear equations. Therefore, the linear equations of the model in Figure 2 can be written as:

(1. Equation) $Z = \beta_1.X + \beta_2.Y + 1.Res_1$

(2. Equation) $W = \beta_3.X + 1.Res_2$

(3. Equation) $T = \beta_4.Z + \beta_5.Y + \beta_6.W + 1.Res_3$

As shown in Figure 2, there are 3 endogeneous (dependent) variables, and the above three equations are written. The residual term of the first equation named as Res1, the residual term of the second and third equation named as Res2 and Res3 respectively. The coefficients of residual terms are fixed as '1'. β coefficients are free parameters. The variables to the left of the above equations are dependent variables, while the ones to the right are independent variables. There is another point of interest in equations. W and Z variables are dependent variables in the first and second equations while they play an independent role in the third equation. In the group of these equations only the variables X and Y are always independent. This is because X and Y variables are exogeneous variables as seen in Figure 2. And the variables T, W and Z are endogeneous variables. Again as seen in Figure 2., residual error of each latent variable correspond to the residual term of each regression equation in which the latent variable has the dependent variable role. Residual terms mainly represent variance that can not be explained by factor (Kline, 2011). The square root of the variance equals to the standard deviation. This value represents how far or near the distribution of the values in a serial from the mean. If the standard deviation is small, the values are scattered close to the serial average.

Measurement errors of the observed variables that form the latent variable are symbolized by "e" notation. Once again, the superiorities of structural equation modelling are that more than one regression analysis can be conducted at the same time and take into account the measurement errors in the observed variables when doing so. For this reason, it can also be regarded as a simultaneous equation system. The three regression equations described above are equations for the structural model. However, as already mentioned, the structural equation model consists of two parts, the measurement model and the structural model. In the example in Figure 2, there are 15 more regression equations in the measurement model as seen below:

(1. Equation) Question1$= \lambda_1.X+e1$

(2. Equation) Question 2$= \lambda_2.X+e2$

(3. Equation) Question 3$= \lambda_3.X+e3$

(4. Equation) Question 4$= \lambda_4.Y+e4$

(5. Equation) Question 5$= \lambda_5.Y+e5$

(6. Equation) Question 6$= \lambda_6.Y+e6$

(7. Equation) Question 7$= \lambda_7.Z+e7$

(8. Equation) Question 8$= \lambda_8.Z+e8$

(9. Equation) Question 9$= \lambda_9.Z+e9$

(10. Equation) Question 10$= \lambda_{10}.W+e10$

(11. Equation) Question 11$= \lambda_{11}.W+e11$

(12. Equation) Question 12$= \lambda_{12}.W+e12$

(13. Equation) Question 13$= \lambda_{13}.T+e13$

(14. Equation) Question 14$= \lambda_{14}.T+e14$

(15. Equation) Question 15$= \lambda_{15}.T+e15$

In these above equations, each observed variable is a dependent variable. Latent variables are independent variables. This is due to the fact that the observed variables and the latent variables are connected in a reflective way in the measurement model.

2.4. Parameters

There are three kinds of parameters in structural equation models. These are called free, fixed and constrained parameters. Free parameters are parameters for which no value is assigned and whose value is to be estimated. Fixed parameters are parameters with a specific value, such as 0 or 1. Constrained parameters are estimated parameters depending on value of other parameters (Raykov & Marcoulides, 2006). In Figure 2, the estimated total number of parameters in equations in both the structural model and the measurement model is 21. Of these, 6 are the regression coefficients forming the structural model and indicated by the β symbol, 15 are the factor loads of the measurement model and are indicated by the λ symbol. In order to estimate the parameters included in the structural equation model given in Figure 2, it is seen that a total of 21 regression models are working together at the same time. The assumption of regression, which is a prerequisite for each regression, needs to be provided. Therefore, the assumptions that apply to the regression models apply within the structural equation models. These are the assumptions of linearity, normal distribution of error terms (normality), lack of multicollinearity, constant variance of error terms (homoscedasticity), and no relation among error terms (authocorrelation). The assumptions about structural equation modeling will be explained in section 2.6.

2.5. Fit Indices

In the method of structural equation modeling, the measures that assess the compliance of the models with the data are called fit indices or fit statistics. There are many fit indices in the literature. Below there are definitions of the most commonly used of these fit indices. The size of the sample should be considered in the analyzes to be done by the structural equation modeling. Because many of the fit indices are affected by sample size. The minimum sample size that must be used in the structural equation modeling method is at least 10 times the number of parameters that can be estimated in the model (Jayaram, Kannan, & Tan, 2004). In addition, the minimum sample size for structural equation modeling is suggested as 150 (Bentler & Chou, 1987). Some researchers suggest that the sample size for Structural Equation Models should be 200-500, at least 200 (Çelik & Yılmaz, 2013).

CMIN is the likelihood ratio chi-square test. This test shows the correspondence between the proposed model and the actual model and it is most commonly used fit indice. As a result of this test, it is evaluated whether the covariance matrix of the sample with which the model is tested is equal to the population covariance matrix. Furthermore, since this test is a difference test, it is not desirable that chi-square value is significant. The fact that the CMIN / DF ratio is less than 3 and the chi-square value is insignificant indicates that the model's overall fit is within acceptable limits (Meydan & Şen, 2011). The DF, ie the degree of freedom, is calculated from the number of observations in a model and number of the parameters needed estimation. Assumed that the number of observed variable in a models equals p. In this model, the estimated number of parameters can not be more than p(p+1)/2. In this case, the degree of freedom is found as follows (Raykov & Marcoulides, 2006):

DF= p(p+1)/2-(number of parameters in the model)

Models with zero degree of freedom in the structural equation model are called saturated models. Saturated model shows perfect fit with data. Negative degree of freedom indicates that the model can not be defined. The model can be defined if the degree of freedom is not negative, ie zero or positive.

Table 2. Good Fit Values

Fit Indices	Goodness of Fit Values
CMIN/DF	0 <CMIN/DF< 2
CFI	0,97 < CFI < 1
AGFI	0,90 < AGFI < 1
GFI	0,95 < GFI < 1
NFI	0,95 <NFI< 1
RMSEA	0 < RMSEA < 0,05

Source: Bayram, N. (2013). Yapısal Eşitlik Modellemesine Giriş. Bursa: Ezgi Kitapevi.

CFI (Comparative Fit Index) is a fit indice that compares the saturated model with the independent model. In the independent model, there is no relationship among the dimensions that form the research model. CFI values can range from 0 to 1, values above 0,90 and close to 1 show good fit (Schermelleh-Engel, Moosbrugger, & Müller, 2003). CFI is in the group of fit indices based on independent models.

Table 2 summarizes the goodness of fit values according to the literature. Table 3 summarizes acceptable fit values according to the literature. The AGFI fit indice is calculated using the degree of freedom. It is affected by sample size. When the sample size increases, the value of the AGFI indice also increases. AGFI takes a value between 0 and 1. Values over 0.90 indicate that the fit is good (Bayram, 2013).

Table 3. Acceptable Fit Values

Fit Indices	Goodness of Fit Values
CMIN/DF	2 <CMIN/DF< 3
CFI	0,95 < CFI < 0,97
AGFI	0,85 < AGFI < 0,90
GFI	0,90 < GFI < 0,95
NFI	0,90 <NFI< 0,95
RMSEA	0,05 < RMSEA < 0,08

Source: Bayram, N. (2013). *Yapısal Eşitlik Modellemesine Giriş. Bursa: Ezgi Kitapevi.*

The GFI fit indice is a measure of the degree of variance and covariance that is explained by the model. The value of the GFI fit indice rises as the sample size increases. This feature can prevent accurate results when sample size is low. The GFI value ranges from 0 to 1. Values above 0.90 are considered acceptable model indices. Values above 0.90 indicate that covariance is calculated among the observed variables. GFI and AGFI fit indices are based on the residuals (Bayram, 2013).

RMSEA is a measure of fit that compares the mean differences of each expected degree of freedom that can occur in the population with each other. This scale is adversely affected by sample size. A value of 0.05 or less for the RMSEA fit indice indicates good fit (Bayram, 2013). Values between 0.05 and 0.08 indicate acceptable fit (Byrne, 2010).

NFI (Normed Fit Index) takes values between 0 and 1. Higher values show better fit. Values greater than 0.90 are acceptable, while values greater than 0.95 are good fit. It is in the group of the fit indices based on independent model.

In the "output model fit" section of the AMOS program, the model's fit indices are displayed as follows. In the following example, the values of CMIN/DF and RMSEA are acceptable. However other fit indices show a problem. In this case, indices may come up to normal values after model modifications. Fit indice values should be read from the "default model" line. Default model refers to the model being tested. On the bottom line there is the "saturated model". Saturated model is the mode where the degree of freedom is zero and the data is perfectly matched to the model. For this reason, the indice values of the best model are in this line. On the bottom line there is independence model. It's the worst possible model. This line contains the worst possible indice values. The CMIN (chi-square likelihood ratio) value appears to be significant when the P value is 0,000 in the example below. However, this test is required to be insignificant since it is a difference test. However, in most cases this value is significant. This may be due to the neglect of some of the assumptions of structural equation modeling described in the next section as general practice during analyzes.

Model	NPAR	CMIN	DF	P	CMIN/DF
Default model	111	1529,801	555	,000	**2,756**
Saturated model	666	,000	0		
Independence model	36	9511,261	630	,000	15,097

Model	RMR	GFI	AGFI	PGFI
Default model	,045	**,827**	**,793**	,689
Saturated model	,000	1,000		
Independence model	,248	,211	,166	,199

Model	NFI Delta1	RFI rho1	IFI Delta2	TLI rho2	CFI
Default model	,839	,817	,891	,875	**,890**
Saturated model	1,000		1,000		1,000
Independence model	,000	,000	,000	,000	,000

Model	RMSEA	LO 90	HI 90	PCLOSE
Default model	**,066**	,062	,070	,000
Independence model	,186	,183	,190	,000

In addition to the fit indices described above, there are also indices used for model comparison. These are called model comparison adaptation indices. AIC (Akaike Information Criterion) is one of them. In the compared models the one which has lowest AIC value is considered as the closest model to reality (Karagöz, 2016).

2.6. Assumptions of Structural Equation Modeling

Similar to regression analysis, structural equation modeling has its assumptions. But in structural equality models, many regression equations work together, whether in the structural model part or in the measurement model part. Therefore, the assumptions that apply to the regression models are valid for the structural equation models. As these assumptions are known, linearity, that is, the relationship between dependent and independent variables is linear, normal distribution of error terms (normality), no multicollinearity which means independent variables are not related to each other, the variance of error terms is fixed (homoscedasticity) or in other words there is no relationship between independent variables and error terms ve no authocorrelation that means that there is no relationship between error terms (Wooldridge, 2003). If these assumptions are met, it should be considered whether the assumptions required for the structural equation models are also met. These assumptions can be summarized as follows (Bayram, 2013).

- **Observed variables have multivariate normality:**

The multivariate normal distribution is the most important assumption of the maximum likelihood estimation method used in structural equation modeling. This rule is often violated when ordinal and discrete scales are used. Neglecting the assumption of multivariate normal distribution of observed variables leads to a high CMIN / DF value and a significant test outcome. In case of violation of this assumption, it is recommended to use the estimation methods such as weighted least squares (WLS) instead of the maximum likelihood estimation method. This method can be used if the data is continuous but does not meet the normal distribution requirement. Other prediction methods that may be preferred in this case are ADF (asymptotically distribution free), MLM (Robust Maximum Likelihood) and GLS (generalized least squares) (Tabachnick & Fidell, 2001). As the complexity level of the model tested in the structural equation modeling method increases, the number of sample observations must also be increased. However, as the distribution of the data becomes farther away from the normal distribution, it is necessary to increase the number of data (Kline, 2011).

The skewness and kurtosis values are examined to determine whether the variables in the data set are normally distributed. These values are calculated on the basis of moments. In general, the packaged softwares calculate these values to be 0 as base value. In this case values between -2 and +2 are considered normal. In addition, Kolmogorov-Smirnov and Shapiro-Wilk tests can be conducted to test whether the data set is normally distributed (Sarstedt & Mooi, 2014).

In cases where the data set does not fit the normal distribution, the outliers (extreme values) should be cleared first. In AMOS program in analysis properties window in output tab, normality and outliers can be tested. It is sufficient to mark the "test for normality" and "outliers" options so that these test values can be obtained in a tabular form. Additionally in SPSS, outliers can be determined by examining the Mahalonobis distance value. If the dataset is not normally distributed, what can be done is covered in the last section of this book.

- **Latent variables have multivariate normal distribution:**

It refers to the endogeneous latent variables have normal distribution. In practice, it is a violated assuption.

- **Linearity:**

As stated at the beginning of the book, structural equation modeling is a component of factor and regression analysis. Therefore, linearity, which is the most important assumption of regression analysis, also applies to structural equation modeling. In the structural equation model, it is assumed that there are linear relationships between latent variables and also between observed and latent variables.

- **Absence of outliers:**

The outlier affects the significance of the existence model negatively.

- **Multiple measurements:**

In the structural equation model, three or more observed variables must be used to measure each latent variable.

- **No multicollinearity:**

It is assumed that there is no relation between the independent variables in the structural equation model.

- **Sample size:**

In the structural equation modeling, many of the fit indices are influenced by sample size. In some sources, a minimum sample size of 150 is recommended for structural equation models (Bentler & Chou, 1987). The minimum sample size that should be used in the structural equation modeling method is at least 10 times the number of parameters that can be estimated in the model. (Jayaram, Kannan, & Tan, 2004). According to some researchers, the sample size required for structural equation modeling should be at least 200 and 200-500 (Çelik & Yılmaz, 2013).

- **No correlation between error terms:**

It is assumed that there is no correlation between error terms in the structural equation modeling method. However, if it is explicitly stated by the researcher in the conceptual model, a correlation can be made between the error terms (Doğan, 2015).

2.7. Types of Structural Equation Models

There are four basic types of structural equation models. These are explained below:

2.7.1. Path Analysis Models

In the method of structural equation modeling, the models established with only observed variables are called path analysis models. The basis of the structural equation modeling depends upon path analysis. Path analyzes first started to be implemented in the 1920s. Developed by biologist Sewall Wright (Taşkın & Akat, 2010). The path analysis is similar to multiple regression as it is done with observed variables. But it is superior than multiple regression. Because there is one dependent variable in the multiple regression. However, there may be more than one dependent variable in the path analysis, and a variable can be both a dependent variable and an independent variable. In path analyzes, more than one regression model can be analyzed at the same time, and indirect and direct effects can be measured at the same time. Direct effect is the effect of one variable on another variable without any mediation. However, the indirect effect arises from the intervention of a variable which is playing mediator role between independent and dependent variables. This variable is named as the mediator variable. The sum of the direct effect and the indirect effect of a variable on another variable is called the total effect (Raykov & Marcoulides, 2006). Since path analyzes do not contain latent variables, they can not be saved from measurement errors (Meydan & Şen, 2011). For this reason, structural regression models generated by latent variables give more accurate results. Because structural regression models include measurement model. Path analysis models and examples will be discussed in more detail in chapter 5.

2.7.2. Confirmatory Factor Analysis Models

Factor analysis is divided into two types as exploratory and confirmatory. In explanatory factor analysis, factors are revealed from relations among variables. In explanatory factor analysis, the observed variables can be loaded on any factor or on multiple factors. However, in the confirmatory factor analysis, the theoretically predetermined factor structure is confirmed by the current data. In other words, in the confirmatory factor analysis, which factor will be loaded on an observed variable is predetermined. By means of the explanatory factor analysis, the latent variables are revealed from the observed variables. However, in the confirmatory factor analysis, previously discovered scales are confirmed again with the collected data. In Section 3, examples of first-level one-factor, first-level multi-factor and second-level multi-factor confirmatory factor analysis are included.

2.7.3. Structural Regression Models

It is regression models formed between latent variables in structural equation models. It consists of a combination of measurement model and structural model. Incorporating the measurement model and the structural model allows the inclusion of measurement errors so that more accurate results can be obtained. In other words, confirmatory factor analysis and multiple regression analysis coexist. Structural regression models will be discussed in more detail in Section 5.

2.7.4. Latent Change Models

They are also named as "latent change models", "latent growth curve models" or "latent curve analysis". Models that describe longitudinal variation in time series (Raykov & Marcoulides, 2006). These models are the models used to explain the growth and decay of an event over time, similarities or differences within and between units. (Doğan, 2015). In Figure 4, two factorial growth models are observed for two time points (T1, T2). Structural equation modeling is a very useful method for analyzing changes in time. Repeated measurements over time are needed to use the latent change models. Such data are called longitudinal data (vertical cross-section data). In AMOS program, under the "Plugins" menu, latent change models can be drawn from the "Growth Curve Model" window. According to Baltes and Nesselroade, this model can be used for the following purposes. (Baltes & Nesselroade, 1979):

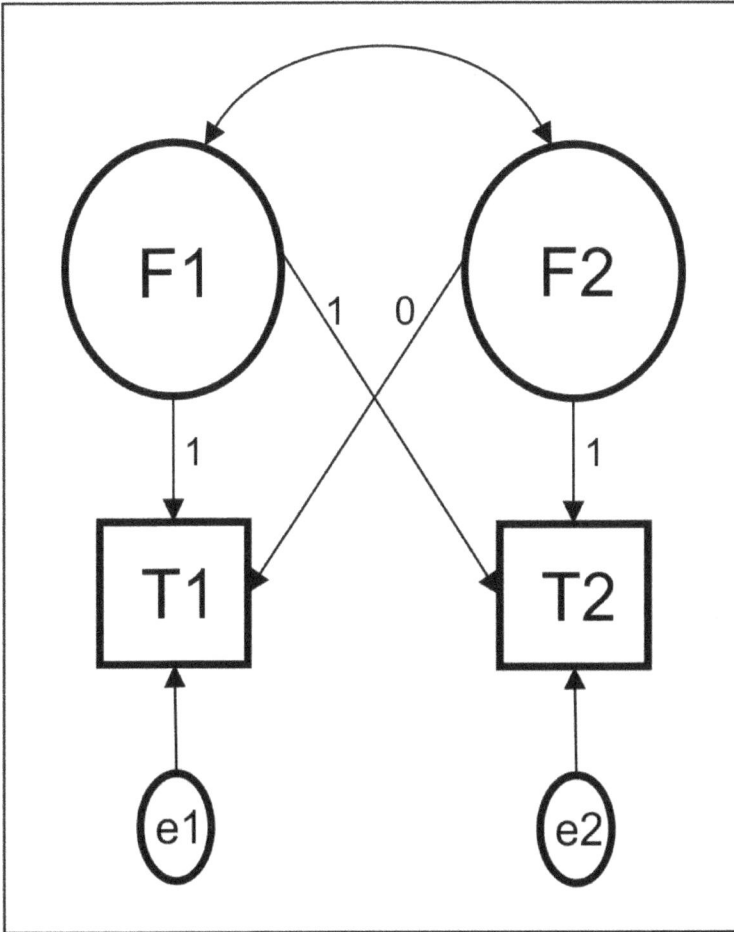

Figure 4. Latent Change Model Example

Source: Raykov, T., & Marcoulides, G. (2006). A First Course in Structural Equation Modeling. Mahwah: Lawrence Erlbaum Associates.

(1) Describe observed and unobserved vertical section data.

(2) Characterize the development of individuals and groups.

(3) To predict individual and group differences in developmental forms.

(4) To examine the dynamic determinants among variables in time.

(5) To reveal the group differences of the dynamic determinants between variables in time.

3. VALIDITY AND RELIABILITY ANALYSIS

Scale is the method used to find the numerical values of the dimensions that constitute a concept. Since concepts can not be directly measured in social sciences, questionnaires are formed to define these concepts. Reliability means that a scale is always measure the same value under the same conditions consistently. For example, a questionnaire form is reliable if the same group is given the same result when applied two different times. So if we ask the same questions about the same people, if the conditions are not changed, they are expected to give the same answers. Otherwise, this means that the persons in the sample either they did not understand the questions on the questionnaire or they did not read them. Validity is a measure of what we really want to measure. For example, if a questionnaire actually measures a different concept than the dimension we want to measure, it is not valid. If the questions we ask about the concept A are confused with the questions about the concept B, then it means that the concepts we consider to measure are not perceived or perceived as different from those in the sample. In this case, the scale we use is not a valid measurement tool for this sample. For this reason, it is necessary to test the validity and reliability of the scale before any analysis is started. As a result of these tests, verification of unidimensionality is generally provided. Unidimensionality means that the observed variables used to measure each dimension must measure only one dimension (Avcılar & Varinli, 2013). Construct validity and reliability must be determined in order to confirm unidimensionality. In a theoretically determined model, construct validity refers to convergence of observed variables that are connected to the same latent variable (convergent validity) and dissociation of observed variables from other observed variables that are connected to other latent variables (discriminant validity). The construct validity indicates that the observed variables do not measure any latent variable other than they connected in the conceptual model. But in this case it would not be correct to say that the validity of the construct is fully realized without confirming the reliability of the scale (Gerbing & Anderson, 1988).

3.1. Determination of Convergent Validity

Convergent validity indicates that the correlations between questions constituting a construct are high. In structural equation modeling method, it is necessary to look at the results of confirmatory factor analysis to determine the convergent validity of the scales used to measure the dimensions constituting the conceptual model of the research. The measurement model part of structural equation models correspond to confirmatory factor analysis (Confirmatory Factor Analysis - CFA). Therefore, if the measurement model fit indices are low, there is no need to test the structural model (See also Figure 1. Demarcation between Measurement Model and Structural Model). Because the scales used to measure the dimensions that make up the conceptual model will not be validated. Therefore, if the measurement model is insufficient, the fit indices of the structural model will be low. The t test results of all the coefficients in the measurement model should indicate that the coefficient values are different from zero. The standard value of each coefficient in the measurement model is the factor loadings of the confirmatory factor analysis. Each factor load should be higher than 0.50. Otherwise, the fit indices of the general model will be adversely affected. The fact that the factor loads are above 0,5 is evidence of convergent validity. If the critical rate value of a question in CFA results is greater than 2 as an absolute value this means that this item is loaded to the factor it is connected.

In structural equation modeling method, while the CFA model is set, some parameters are freed while some are fixed. As shown in Figure 5, Parameter of question 1 and parameters of error terms are fixed by assigning 1 value. Others are fixed. The freed parameters will be estimated by the program.

Before applying confirmatory factor analysis (CFA), it is first necessary to look at the results of explanatory factor analysis (EFA) in practice. Even though scales generally accepted in the literature are used, to see if the survey fillers correctly perceive the questions principle component analysis should be conducted in SPSS before set up CFA model in AMOS. And how many different dimensions the questions are perceived by those who solve the questionnaire should be clarified.

At this stage, the necessary questions should be eliminated. This step is also called the purification stage. Principle component analysis is a type of analysis that assigns the variables in the data set into groups so that the relationship between the variables in the group is maximized. Main purpose of this analysis is to obtain the least number of factors to represent the relationship among items at the highest level.

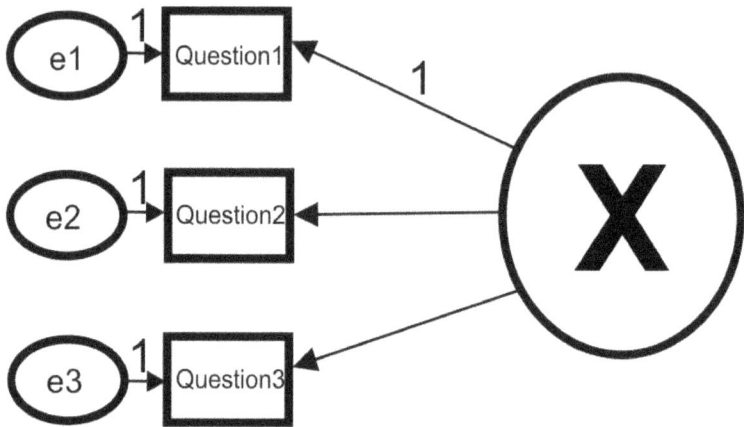

Figure 5. Single Factor CFA Model

Bartlett test of Sphericity and Kaiser-Meyer-Olkin test were performed to determine the suitability of the dataset to the principle component analysis. If the null hypothesis is not rejected as the result of the Bartlett test of Sphericity, the analysis is not continued. This test detects whether the correlation matrix indicating the inter-variable relation is a unit matrix. There is no relation between variables in case of unit matrix. The Kaiser-Meyer-Olkin test is affected by sample size. This test compares the values of the observed correlation coefficients with the values of the partial correlation coefficients. In this way, it tests whether the sample size is sufficient to perform principle component analysis. Values above 0.7 are considered good. If it is below 0.5, factor analysis can not be continued (Karagöz, 2016).

Essentially, principle component analysis is done to determine scale validity, but it also fulfills the purpose of making the data set analyzable. It fulfills the following functions for this (Aksu, Eser, & Güzeller, 2017):

- To remove the dependency between variables.

- To obtain fewer new variables those are not related to each other.

Table 4. Results of Confirmatory Factor Analysis

Items	Conceptual Variable	Standardized Factor Loads	Unstandardized Factor Loads	Standard Error	t-Value (Critical Ratio)
Qestion1		0,818	1		
Qestion2	X	0,906	1,104	0,049	22,523
Qestion3		0,907	1,111	0,049	22,570
Qestion4		0,825	1		
Qestion5	Y	0,732	0,882	0,057	15,549
Qestion6		0,718	0,885	0,058	15,187
Qestion7		0,757	1		
Qestion8	Z	0,835	1,102	0,062	17,785
Qestion9		0,939	1,255	0,062	20,176
Qestion10		0,676	1		
Qestion11	W	0,799	1,131	0,083	13,555
Qestion12		0,785	1,158	0,087	13,379

Note: For all values P<0.01

After conducting explanatory factor analysis and purification by principle component analysis, the remaining indicators (questions) are linked under the structures they belong to and the CFA model is created in the AMOS program as shown in Figure 5 (The principle component analysis performed in the SPSS program is not explained because it is out of scope of this book). The number of factors is freed in EFA. In CFA, the number of factors and which indicators are connected is determined in advance.

Table 4 shows the way in which confirmatory factor analysis results are given. What is important here is that the standard factor loads of the questions under each conceptual variable are over 0.50. By looking at this table, questions with a standard factor load of less than 0.50 are discarded.

In this case factor analysis is done again. Figure 8 shows a second-order CFA model. It is important to note here that residual terms are to be placed in the first-level latent variables (Res1, Res2 and Res3). Figure 9 shows another second-level multi-factor CFA model. The difference of this model is that there are more than one factor in the second level. Since there are more than one exogeneous variables, covariance is placed between the second level latent variables (between A and B). The fit indices of the CFA model are then tabulated. Figure 6 shows the analysis features that must be marked in the Analysis Properties window before running the CFA model in the AMOS program. Figure 7 shows the first-order multi-factor CFA model. In this case it must be considered to place covariance between the latent variables. The rule of placing covariance between exogeneous variables is also valid here. In

Figure 8, covariance is not placed when there is only one exogeneous variable. However Figure 9 shows covariance between two exogeneous variables.

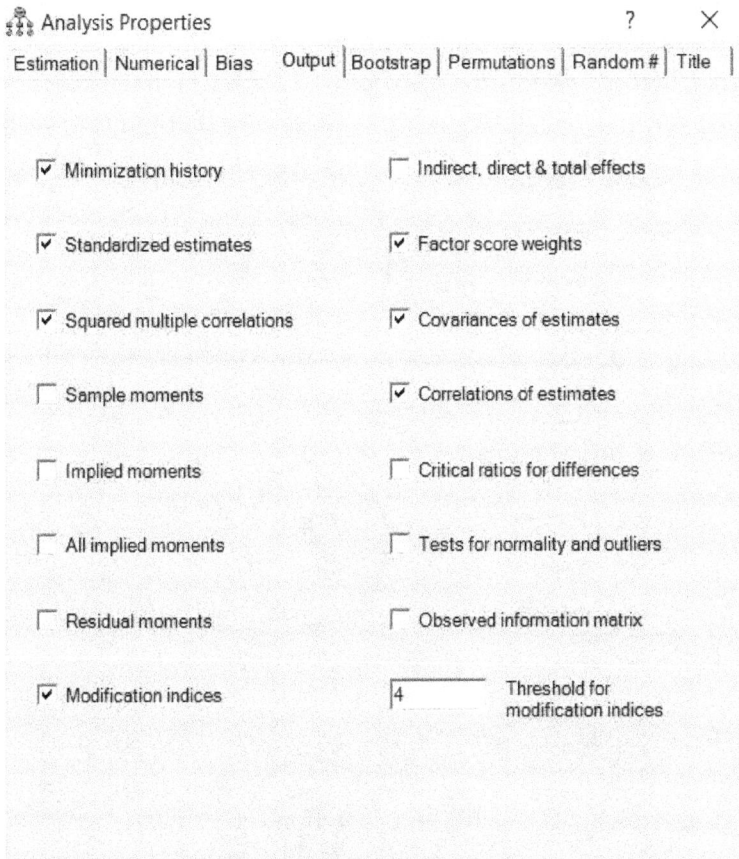

Figure 6. Analysis Properties Window for CFA Models in AMOS program.

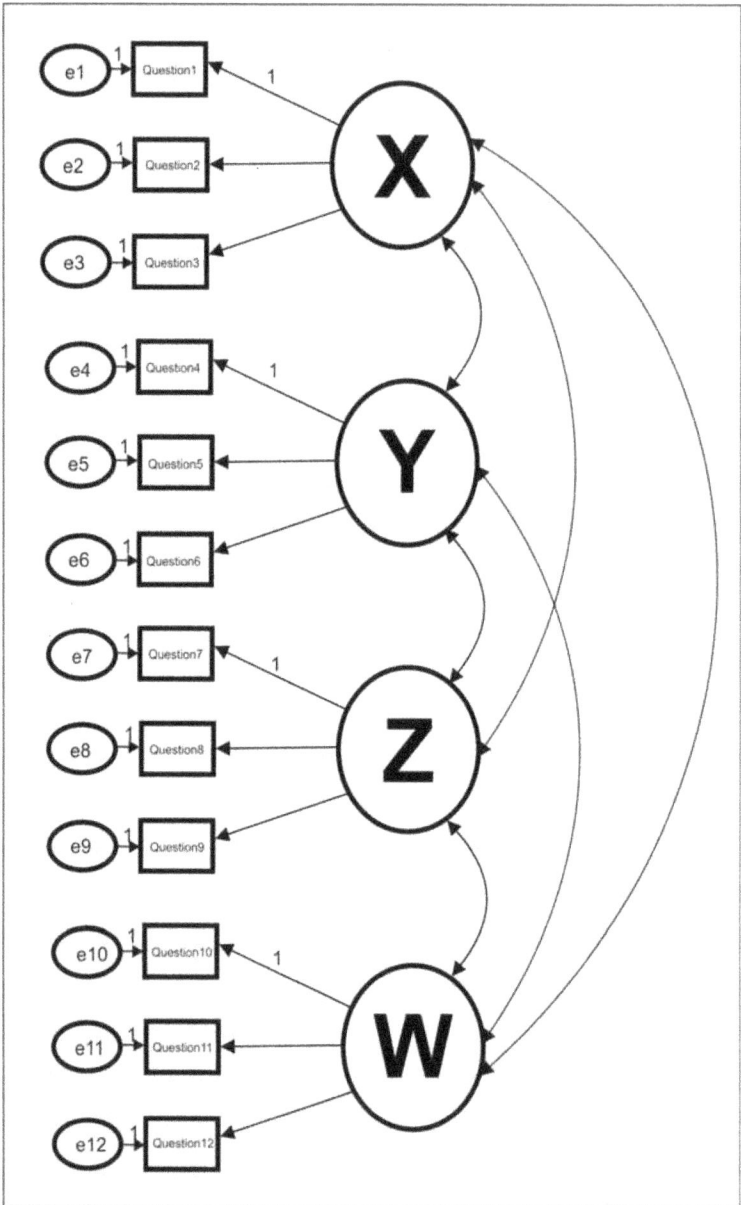

Figure 7. Multi-Factor First-Order CFA Model

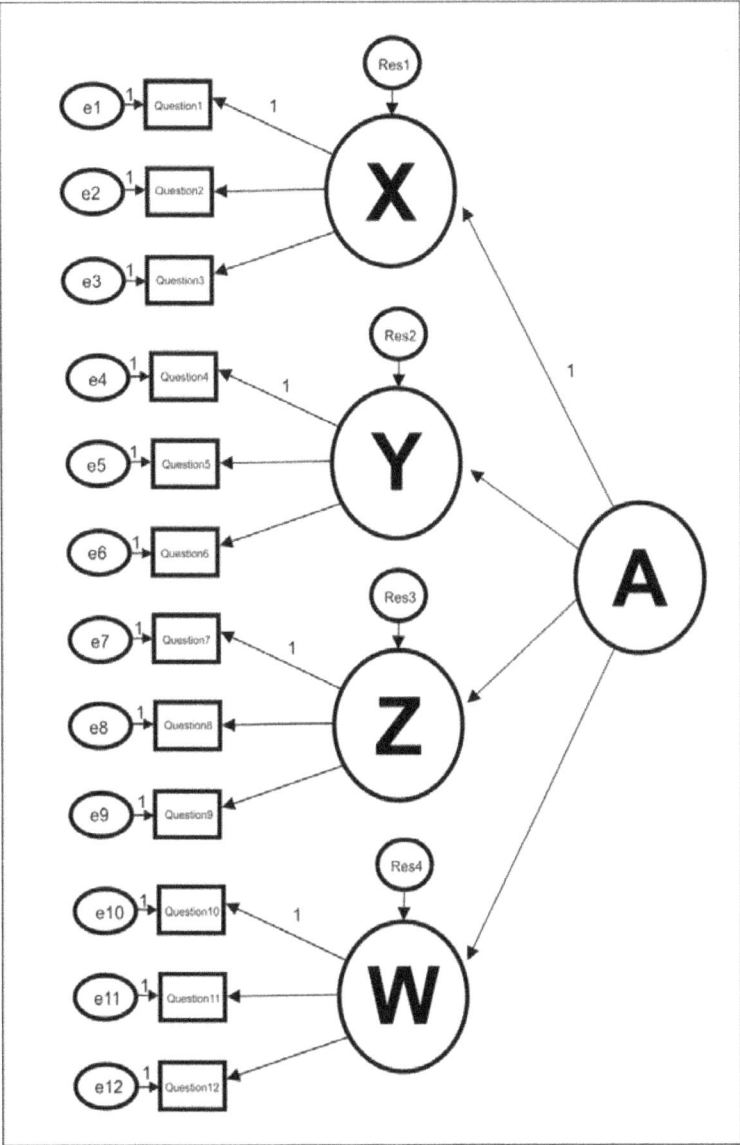

Figure 8. Second-Order CFA Model

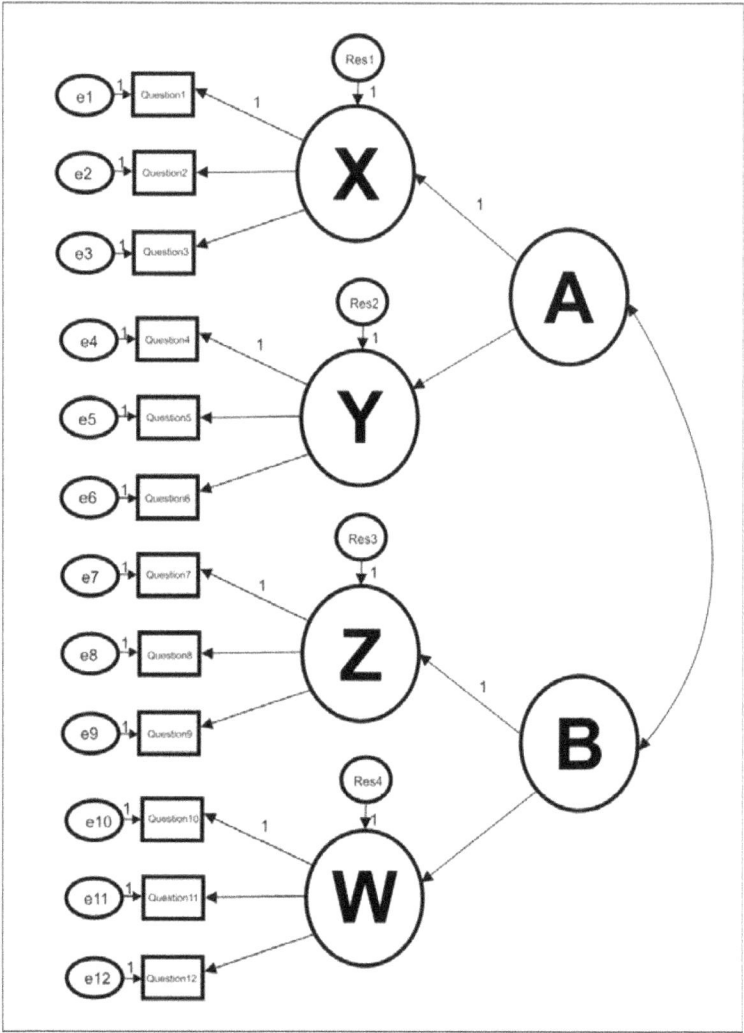

Figure 9. Multi-Factor Second-Order CFA Model

Another indicator of convergent validity is the Average Variance Extracted (AVE) value. To be able to confirm the convergent validity, it must be more than 0.50 or 0.50. (Fornell & Larcker, 1981).

3.2. Determination of Discriminant Validity

Discriminant validity is the measure of the level at which a structure in a measurement model differs from other structures. It is an indicator of a low correlation between the questions that form a construct and other questions that form other construct. To find the discriminant validity for each dimension, we first need to calculate the Average Variance Extracted (AVE) value for each dimension. The acceptable AVE value must be greater than 0.50 or 0.50. However, as noted in the previous section, this value confirms convergent validity when examined alone (Fornell & Larcker, 1981). In order to determine discriminant validity, it is also desirable that the values of the AVE for each construct in the data set are larger than the correlation coefficients of that construct with the other constructs. In this case, it can be determined that the scales used have discriminant validity for each dimension. AVE value alone does not indicate discriminant validity but the square root of the AVE value of each construct is larger than the inter-dimensional correlation value it can be said that there is discriminant validity (Fornell & Larcker, 1981). In Table 5, the values shown in parentheses as crosswise is the square root of the AVE values. For the 9 dimensions in the table, the square root of the AVE values in each column is higher than the correlation coefficients in that column. In addition, the AVE values are above the 0.50 threshold. In this case, it can be said that the scales used for this example have discriminant validity. The AVE value is not calculated by the AMOS package program. However, it is easy to find ready-made excel files that provide this value calculation on the internet.

3.3. Determination of Reliability

After determination of the validity of the scales by means of CFA reliability analysis must be conducted for each construct. First of all, Cronbach's α value is calculated for each dimension separately. Values greater than 0.7 threshold indicate that the internal reliability of the scale used is sufficient. Cronbach's α is a measure based on correlations between items in a consruct. It is obtained by dividing the sum of the variances of the items constituting a scale by the general variance. It takes a value between 0 and 1. Values beyond 0.7 threshold indicate that the scale is reliable. If it is below 0.6, the reliability of the scale is low. (Karagöz, 2016).

Another value that is used to calculate the reliability of the scale for each dimension is the composite reliability value. The composite reliability value is calculated from the factor loads found in the confirmatory factor analysis. After CR values beyond 0.7 threshold or equals to 0.7 it can be said that there is composite reliability (Raykov, 1997).

Table 5 shows a sample table showing Cronbach's α, AVE and CR values calculated for each construct and the correlation values between constructs. Cronbach's α value can be calculated from the scale reliability menu in the SPSS program. The AVE and CR values are found by placing the results of the CFA factor loadings in to the formulas. There are ready-made calculation tools on the Internet.

Table 5. Descriptive Statistics, Correlation Coefficient, Reliability Results and Discriminant Validity

	Avr.	Std. Dev.	1	2	3	4	5	6	7	8	9
1. Construct	3,25	0,81	(0,842)								
2. Construct	3,28	0,71	,216*	(0,711)							
3. Construct	3,63	0,72	,427*	,383*	(0,840)						
4. Construct	3,72	0,68	,228*	,533*	,457*	(0,718)					
5. Construct	3,62	0,70	,449*	,192*	,378*	,298*	(0,769)				
6. Construct	3,76	0,68	,430*	,394*	,551*	,450*	,499*	(0,734)			
7. Construct	3,23	0,87	,585*	,166*	,452*	,174*	,449*	,479*	(0,800)		
8. Construct	3,68	0,67	,394*	,496*	,672*	,508*	,350*	,508*	,358*	(0,722)	
9. Construct	3,02	0,77	,340*	,374*	,353*	,335*	,209*	,302*	,219*	,410*	(0,754)
Cronbach Alfa Reliability Coefficient			0,927	0,861	0,901	0,851	0,781	0,771	0,828	0,808	0,721
Composite Reliability Coefficient (CR)			0,924	0,854	0,905	0,841	0,791	0,777	0,840	0,813	0,725
Average Variance Extracted (AVE)			0,710	0,506	0,706	0,516	0,592	0,539	0,640	0,522	0,570

* $P<0,05$, Note: the values written in brackets indicate the square root of the AVE values.

There are statistically significant relationships among the constructs in the sample in Table 5. Correlation is the coefficient that indicates the power of linear relationship between variables. This coefficient must be statistically significant in order to be able to say that there is a relationship between variables. The correlation coefficient takes a value between -1 and +1 (Sipahi, Yurtkoru, & Çinko, 2010).

4. STRUCTURAL REGRESSION MODEL

As a result of the processes described above, convergence validity and discriminant validity are determined. Then the phase for forming the structural model begins. The structural model is based on measurement model. In this phase it should be noted that the structure and order of CFA model is preserved in the model that will be established with latent variables. Depending on the causal relationships between latent variables, the directions of the arrows are determined in accordance with the developed hypotheses and a structural model is constructed. Two-way arrows indicate the covariance between two variables without specifying the direction of causality. While constructing the structural model, the conceptual model leads the way. The conceptual model is mainly based on the relations found in the literature. Finally, the conceptual model is tested using real data. Before the hypothesis tests, the fit indices of the model are examined.

If the fit indices of the model are not within the limits recommended in the literature, the modifications are made and the fit indices are improved provided that they are compatible with the literature. If the fit indices are at an adequate level, the predicted values of the model parameters are checked first. Then the hypothesis test results of the research are given in a table. Figure 10 shows the properties to be marked in the analysis properties menu in the AMOS program when starting the path analysis. Table 6 shows an example of the hypothesis test results. The values in this table are in the estimates section of the output screen of the AMOS program. The notation *** in AMOS output means that P is equal to zero.

Table 6. Hypothesis Test Results Table Example

Relations	Standard Coefficients	Unstandard d Coefficien ts
X → Z	0.533*	0.594*
Y → Z	0.437*	0.638*
Z → W	0.493*	0.377*

*p < 0.05

Figure 11 and Figure 12 show examples of structural models based on the first-level multi-factor DFA model given in Figure 7.

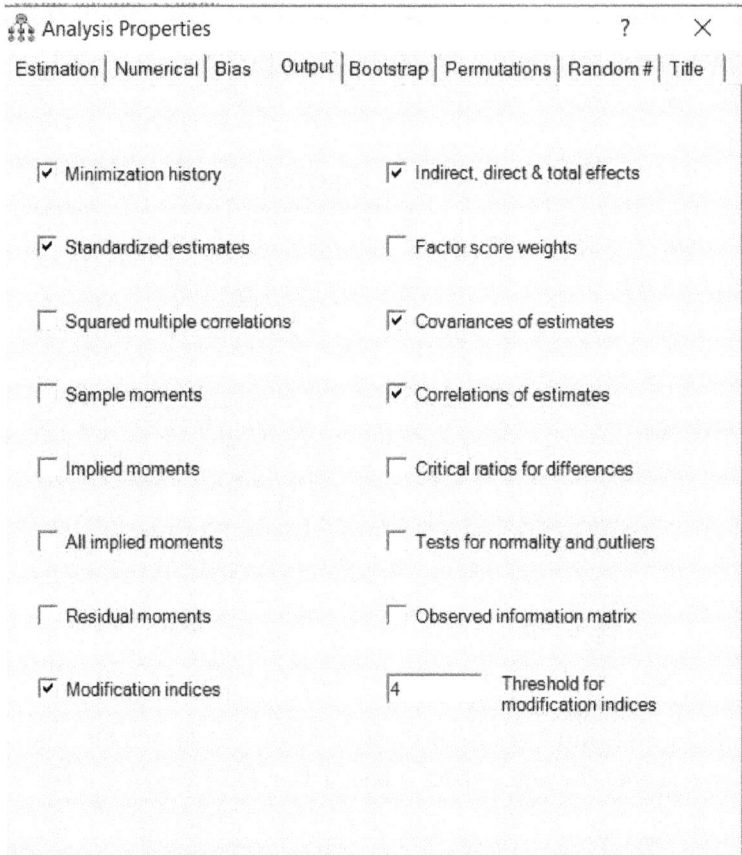

Figure 10. Analysis Properties Window for the Structural Model in the AMOS program.

These examples are called structural regression models. Although the models in Figure 11 and Figure 12 are based on the same measurement model, the path analysis created is different. In Figure 11, there are more than one exogenous variable and therefore covariance is placed between them. In Figure 12 there is only one exogeneous variable. In both models residual terms are linked to the endogeneous variables.

Careful attention should be paid to these rules when constructing structural models. Otherwise the model will not work.

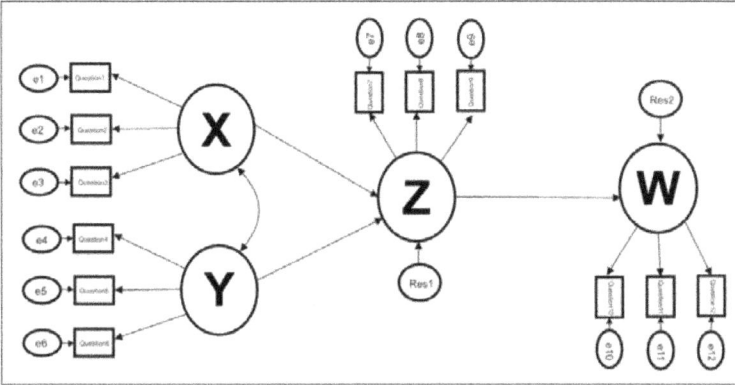

Figure 11. Structural Regression Model Example

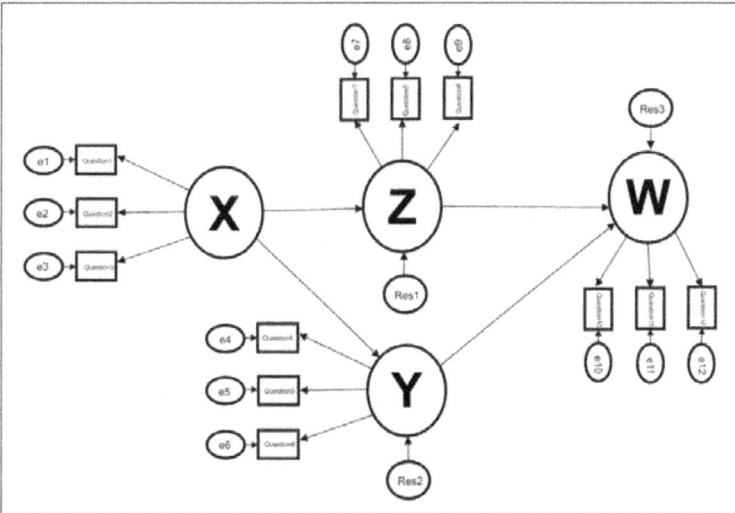

Figure 12. Structural Regression Model Example

5. PATH ANALYSIS

Structural model can be established by directly observed variables in cases factor analysis is done before and the average of the questions that make up the constructs or when working with the secondary data. Such models are called as path analysis. Figures 13 and 14 show the models constructed in Figures 11 and 12, respectively, with the observed variables. As seen in the figures, covariance and residuals remained the same in the models. This is because the rules are the same in the path analyzes made with the observed variables. But the measurement models are gone. Therefore, it is assumed that there are no measurement errors in the path analysis. Predictive error terms are included in dependent variables. Before running the path analysis in the AMOS program, the options in Figure 10 should be checked in the analysis properties window.

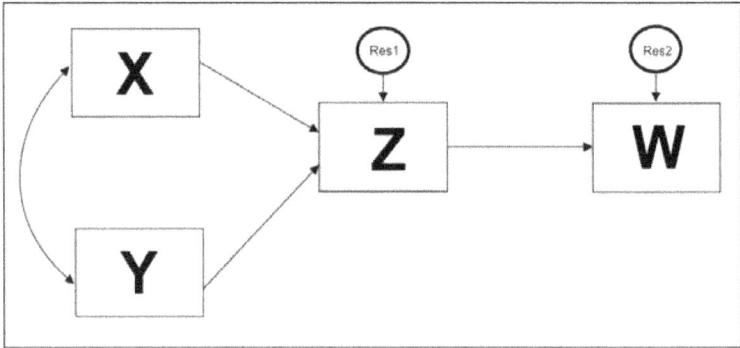

Figure 13. Example of Path Analysis

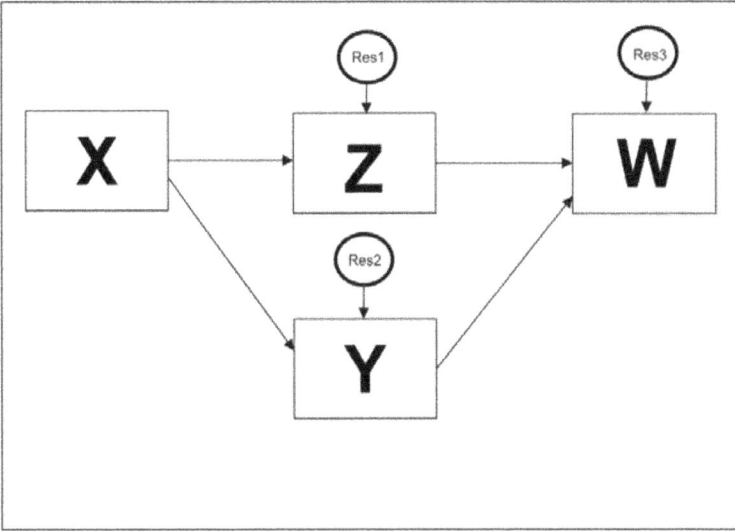

Figure 14. Example of Path Analysis

6. MODEL MODIFICATION

If the fit indices of the structural model do not come to an adequate level, the AMOS program suggests modifications to the user. These modifications improve fit indices. For example, Figure 15 shows the modifications suggested under the Modification Indices menu in the AMOS program. In the case of placing covariance among those with the highest value among the proposed changes in Figure 13, the model's fit indices will improve. Because the modification index value corresponds to the decrease in the chi-square value of the model. However, these covariances are not to be considered as unrelated concepts in the literature that should be taken into account when setting. Because each modification changes the conceptual model which is first introduced. It should be noted that the changes made by this reason do not contradict the purpose of the research and the relations in the literature. To reach the modification indices, the modification indices box in the analysis properties window in the AMOS program must be marked as shown in Figure 10. In the example shown in Figure 15, first the model is run again by adding covariance between the error terms e24 and e22 and compared with the previous situation. Because the highest modification value in the table is 17.457, which is between e24 and e22. After the modifications are made, the model is retested, and if sufficient compliance values can not be obtained, the proposed modifications can be repeated. Again, modifications made should be consistent with the literature. Unrepeatable and inappropriate modifications made are scientifically unacceptable and result in inconsistent situations with the population. For this reason, it is necessary to apply it very carefully. Just in order to raise the fit indices, covariance should not be placed between concepts that are not related to each other.

Covariances: (Group number 1 - Default model)

			M.I.	Par Change
e22	<-->	e21	8,894	-,089
e24	<-->	e21	6,814	-,079
e24	<-->	e22	17,457	,214
e24	<-->	e23	5,889	-,065
e25	<-->	WOM	10,477	,088
e25	<-->	e22	5,651	,100
e25	<-->	e24	6,854	-,112
e19	<-->	e22	4,493	,076
e18	<-->	WOM	7,302	,060
e16	<-->	e20	5,214	,054
e16	<-->	e18	11,396	-,069
e14	<-->	e17	10,965	,155
e14	<-->	e16	12,772	-,092
e13	<-->	e16	5,684	,052
e13	<-->	e15	4,870	-,043
e12	<-->	e16	5,543	,062
e12	<-->	e14	8,850	-,113
e12	<-->	e13	4,865	,072
e11	<-->	e14	4,381	,088
e10	<-->	e24	15,839	,251
e10	<-->	e11	7,053	,160
e7	<-->	e16	4,165	,038
e7	<-->	e15	5,149	-,038
e7	<-->	e14	9,126	-,082
e7	<-->	e13	5,847	,056
e7	<-->	e12	5,577	,065
e6	<-->	e12	5,345	-,055
e5	<-->	e23	4,137	-,036
e5	<-->	e16	7,674	-,057
e5	<-->	e15	5,768	,045
e4	<-->	e22	4,664	-,094
e4	<-->	e17	4,506	,100
e4	<-->	e5	8,962	,089

Figure 15. AMOS Modification Indices

7. MEDIATOR VARIABLE ANALYSIS METHODOLOGY

The variable starting the causality relation between the independent and the dependent variable is called as mediator variable (Wu & Zumbo , 2008). It is also called as intervening variable (MacKinnon, Lockwood, Hoffman, West, & Sheets, 2002). It can also be defined as the variable that transfers the indirect effect of an independent variable to the dependent variable. Analysis of mediator variable is based on the hierarchical regression method introduced by Baron and Kenny in 1986. In order to apply this method, the following conditions must first be met (Baron & Kenny, 1986) :

a. The changes that occur in the independent variable cause a change in the mediator variable,

b. Changes in the mediator variable cause changes in the dependent,

c. If the mediator and independent variables are together included in the regression analysis, the effect of the independent variable on the dependent variable either falls or completely ceases.

In Figure 16, there is a sample mediator variable model. After this model is created, it is first checked whether there is a correlation between all variables. It is thus tested whether the model meets the first two preconditions put forward by Baron and Kenny. Table 7 shows an example of correlation table.

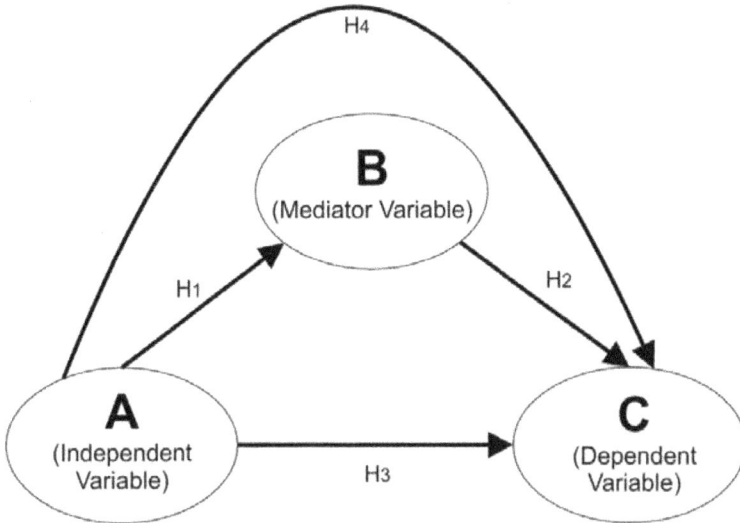

Figure 16. Mediator Variable Model and Hypotheses

Table 7. Example of Correlation Coefficient Table

Constructs	A	B	C
A	-	-	-
B	$0,492^*$	-	-
C	$0,575^*$	$0,672^*$	-

* P< 0.01

When an mediator analysis is performed, three different models are run and the coefficients of the models are compared with one another. Models and hypotheses tested are as follows:

H$_1$: Variable A affects variable B in the positive direction.

H$_2$: Variable B affects variable C in the positive direction.

H$_3$: Variable A affects variable C in the positive direction.

H$_4$: Variable B plays mediator role in the relationship between Variable A and Variable C.

Model 1: $C_i = \beta_0 + \beta_1 A_i + \varepsilon_i$ $\quad\quad\quad$ (H$_3$)

Model 2: $B_i = \beta_0 + \beta_1 A_i + \varepsilon_i$ $\quad\quad\quad$ (H$_1$)

Model 3: $C_i = \beta_0 + \beta_1 A_i + \beta_2 B_i + \varepsilon_i$ $\quad$ (H$_2$ ve H$_4$)

Once the three regression models given above are run separately, the results found are compared as seen in Table 8. As shown in Table 8, when the B variable is added in model 3, the coefficient of the relationship between A and C is lowered and turns to be insignificant. This indicates that the variable B has mediator role in the relationship between variable A and variable C. As a result, hypotheses H$_1$, H$_2$, H$_3$ and H$_4$ are accepted. In this way, the Baron and Kenny method can be easily applied in the SPSS program in the presence of a third variable that plays a role of mediator variable between the two variables.

Table 8. Example of Regression Coefficients Table

Relations	Model 1	Model 2	Model 3
A → C	0,492*	-	0,07
A → B	-	0,575*	-
B → C	-	-	0,582*
R^2	0,242	0,331	0,468
Adjusted R^2	0,240	0,329	0,466
F	129,216*	200,205*	177,925*

* P<0.001

Structural equation modeling can be used in mediator variable analysis when testing multiple mediator variables or multiple independent or dependent variables. For example, in the conceptual model shown in Figure 17, there are two mediator variables.

In the same way, The Baron and Kenny method is applied while the analysis of the mediator variables is done in the structural equality models. For this reason, once the model is constructed, first, it is checked whether there is a correlation among all variables. It is thus tested whether the model meets the first two preconditions put forward by Baron and Kenny. Table 9 shows an example of correlation table. When an mediator analysis is performed, the three different models shown in Figure 18, Figure 19, and Figure 20 below are run in AMOS and the coefficients of the models are compared against each other. Table 10 shows an example of a table comparing the β coefficients found after running models in AMOS.

As shown in Table 10, the coefficient β for the relationship between A and D, which is significant and high in Model 1, has fallen and become insignificant with the inclusion of the variables B and C in Model 3. In this case, the mediator role of B and C are found statistically significant.   In addition, fit indices of each model tested should be given. Therefore Table 10 shows the fit indices of each model.

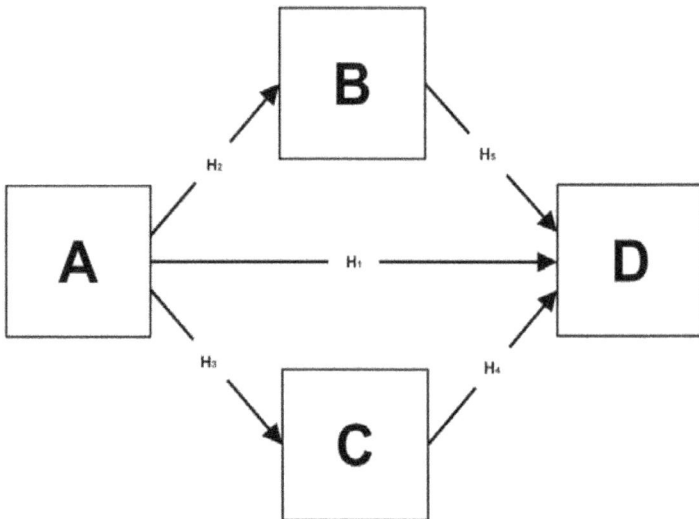

Figure 17. Two Mediator Conceptual Model and Its Hypotheses

For scientific validation of mediator variable roles, these fit indices must also be within acceptable limits. This method can be applied to more variables. The basic logic of this analysis is the comparison of the coefficients of the model to which the mediator variables are not included and the coefficients of the next model to which the mediator variables are included.

Table 9. Correlation Coefficients Sample Table

Variables	1	2	3	4
1. A	—			
2. B	0.883*	—		
3. C	0.861*	0.928*	—	
4. D	0.430*	0.653*	0.703*	—

*p < 0.01.

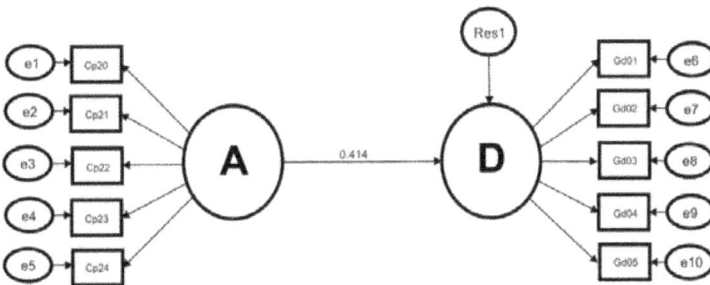

Figure 18. Model 1

If the statistically significant coefficients become insignificant, the role of intermediate variable is confirmed. If the p value of the coefficient remains significant but there is a serious decrease in the coefficient, it can be said that there is semi-mediator effect. In the conceptual model in Figure 17, the mediator roles of B and C can be tested together. But if needed, these variables can be included to analyses one by one.

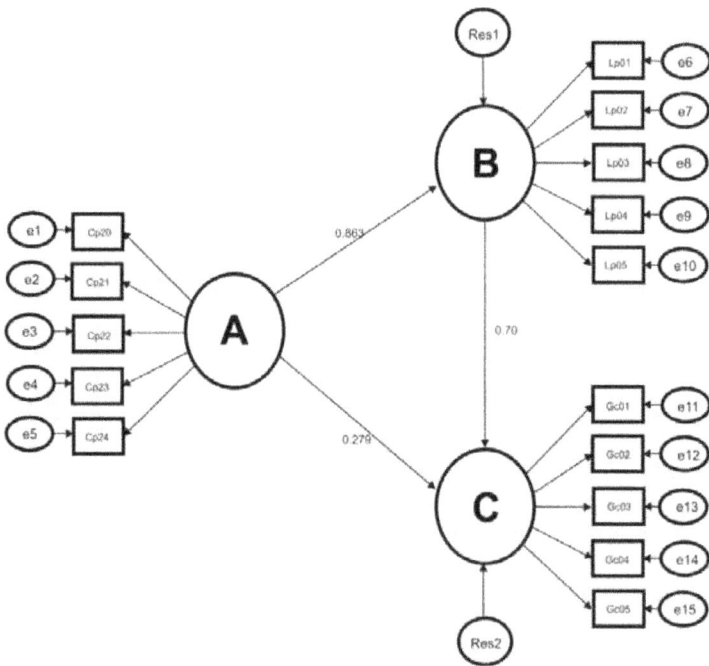

Figure 19. Model 2

Tablo 10. Analysis Results Sample Table

Relations	Model 1	Model 2	Model 3
A → D (H_1)	0.414*		0.035
A → B (H_2)		0.863*	0.865*
A → C (H_3)		0.279*	0.870*
C → D (H_4)			0.532*
B → D (H_5)			0.935*
Model Fit Indices	$\chi2/df=2.554$ CFI=0.985 IFI=0.985 RMSEA=0.14	$\chi2/df=2.921$ CFI=0.954 IFI=0.954 RMSEA=0.14	$\chi2/df=2.856$ CFI=0.947 IFI=0.947 RMSEA=0.14

Note: Path analysis coefficients are standardized.

*$p<0.01$

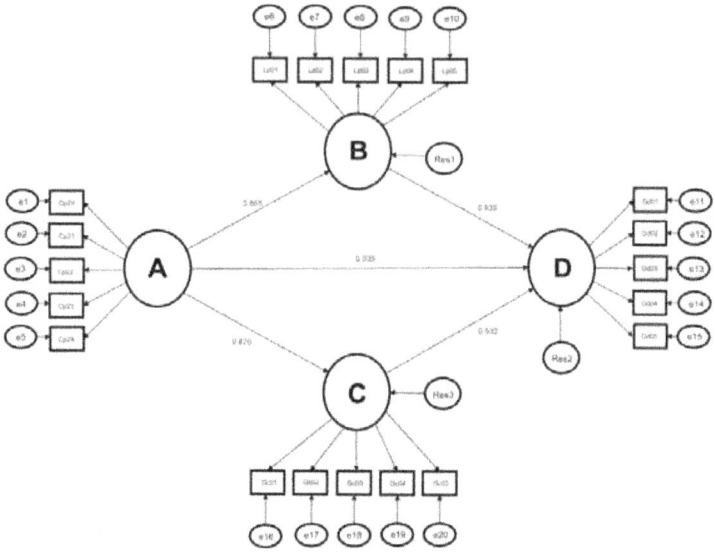

Figure 20. Model 3

II. CHAPTER

APPLICATIONS

8. USE OF AMOS PROGRAM

This section provides general information about the menus and use of the AMOS program. The AMOS program runs on the Windows operating system. It is a graph based program. In this way, users can easily analyze by drawing conceptual models they create. The results of the analysis are given both on the drawn model and also on the tables.

 As the AMOS program carries copy and paste features of the Windows operating system, the results can easily be transferred to programs such as Word and Excel. To open the AMOS program in Windows 10 operating system, the Start → IBM SPSS Statistics → AMOS Graphics menus must be followed. After the AMOS program is turned on, the screen shown in Figure 21 is displayed. There are various icons on the left side of this screen. The functions of these icons are summarized in Figure 22. When you wait for a while on these icons with the mouse, the function of the icon is displayed on the screen. At the top of the screen there are File, Edit, View, Diagram, Analyze, Tools, Plugins and Help menus.

The AMOS program is a program that works together with the SPSS program. For this reason, it is first necessary to contact the SPSS file containing the data to be worked on before starting the model drawing. For this, File → Data Files menus are followed and the window that appears in Figure 23 opens. In this screen, click the File Name menu to select the relevant file and click OK to close the window. In this way, the AMOS file is linked to the data to be run.

When starting to draw the model after this step, firstly the indicator symbol ⬚ is pressed to draw the hidden variable and the three observed variables (indicator) which are shown in Figure 24.

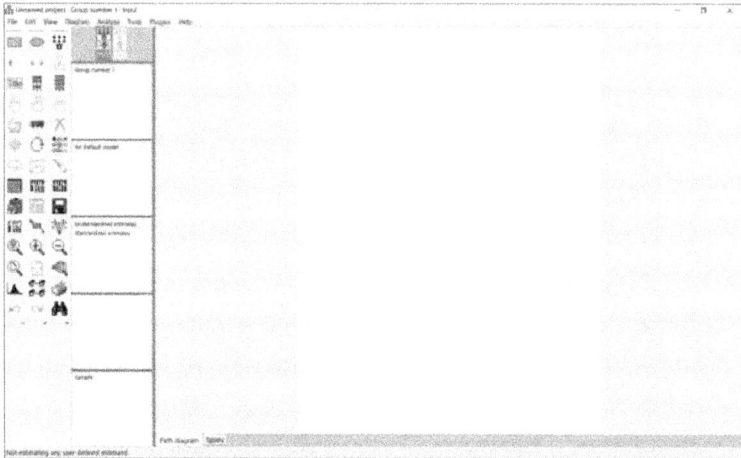

Figure 21. AMOS Startup Screen

Each click creates an indicator. Therefore, it is necessary to click three times to draw three indicators. Then change direction of the indicators by clicking on the direction change icon. One more latent variable is drawn by following the same sequence and the model shown in Figure 25 is obtained. In this model, relations between two latent variables are drawn by using the regression symbol and covariance is drawn between two external variables using the covariance symbol .

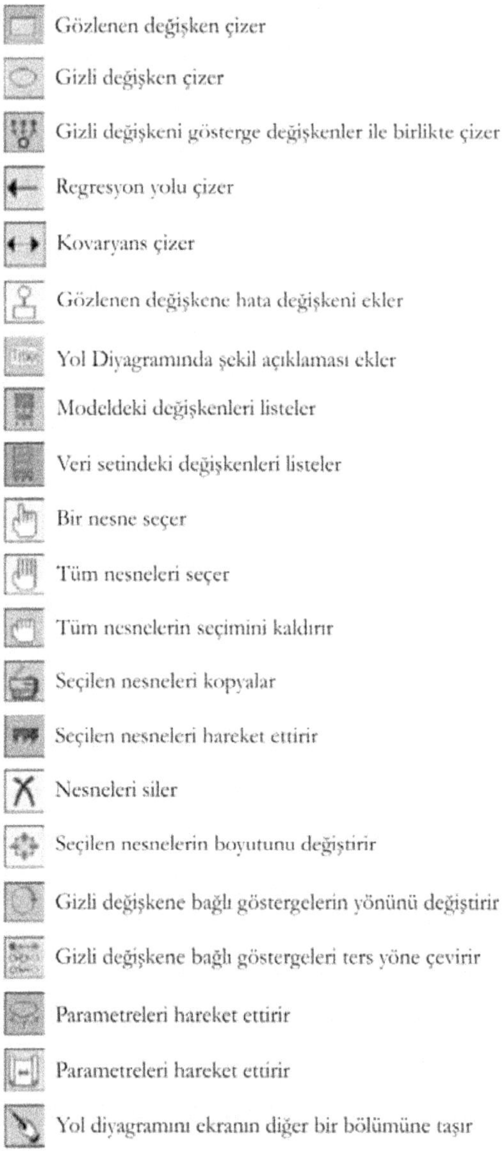

Gözlenen değişken çizer

Gizli değişken çizer

Gizli değişkeni gösterge değişkenler ile birlikte çizer

Regresyon yolu çizer

Kovaryans çizer

Gözlenen değişkene hata değişkeni ekler

Yol Diyagramında şekil açıklaması ekler

Modeldeki değişkenleri listeler

Veri setindeki değişkenleri listeler

Bir nesne seçer

Tüm nesneleri seçer

Tüm nesnelerin seçimini kaldırır

Seçilen nesneleri kopyalar

Seçilen nesneleri hareket ettirir

Nesneleri siler

Seçilen nesnelerin boyutunu değiştirir

Gizli değişkene bağlı göstergelerin yönünü değiştirir

Gizli değişkene bağlı göstergeleri ters yöne çevirir

Parametreleri hareket ettirir

Parametreleri hareket ettirir

Yol diyagramını ekranın diğer bir bölümüne taşır

Figure 22. Drawing Tools in the AMOS Program

Veri dosyasını bağlar

Analiz özelliklerini belirler

Analizleri gerçekleştirir

Yol diyagramını panoya kopyalar

Analiz sonuçlarını gösterir

Kaydeder

Nesne özelliklerini tanımlar

Nesne özelliklerini kopyalar

Seçilen nesne gruplarını simetrik hale getirir

Ekranın seçilen kısmını büyütür

Yakınlaştırır

Uzaklaştırır

Tüm sayfayı gösterir

Modeli sayfaya sığdırır

Ekranda mercek çıkartır

Bayesian istatistiklere dayanan analizi sağlar

Çoklu grup analizi analizi yapar

Seçilen şekli yazdırır

En son yapılan işlemi geri alır

En son geri alınan işlemi tekrar yapar

Modellemeyi özellik araştırmasına dayandırır

Figure 22. Continued

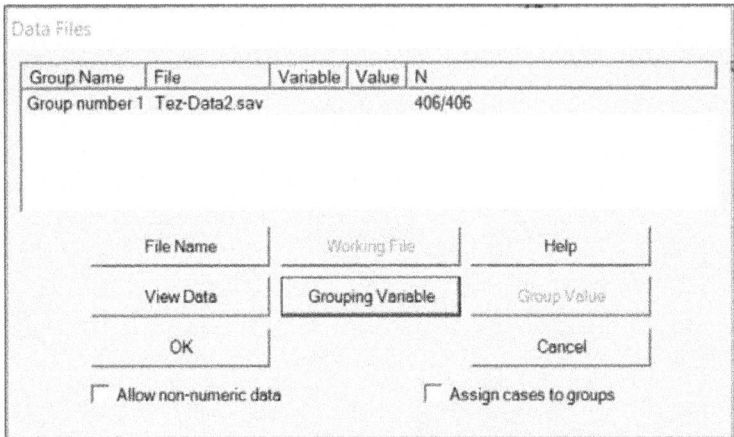

Figure 23. Data Files Window

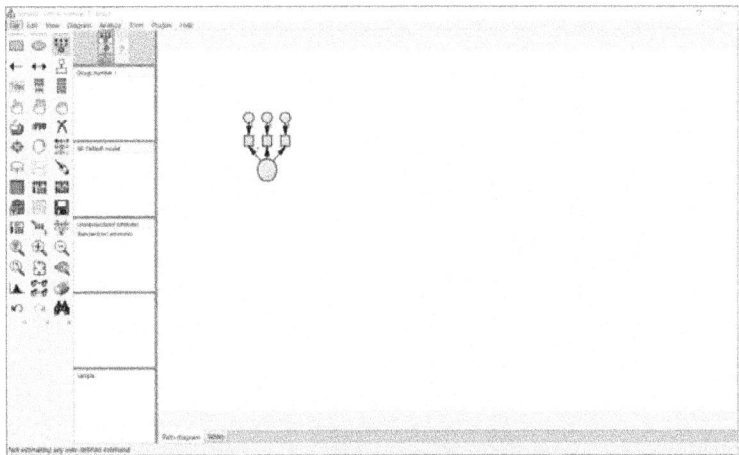

Figure 24. Drawing Latent Variable

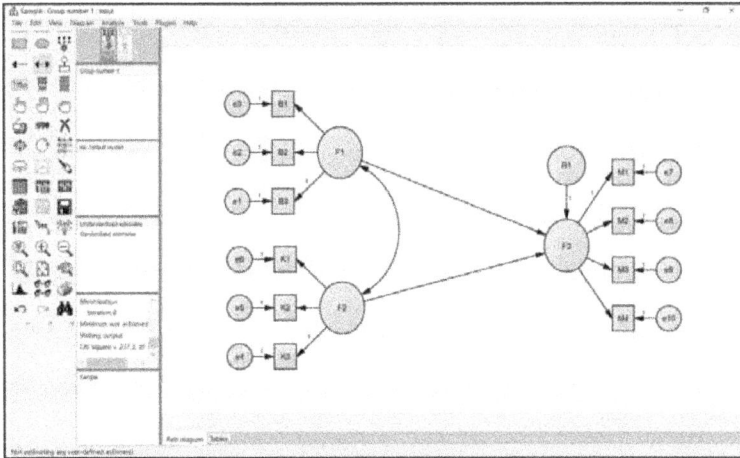

Figure 25. Sample Model Drawing

The data set icon is pressed and the observed variables in the data set are listed. The variables found in this list can be linked to the indicators by means of drag and drop method. Select each latent variable and click on the right mouse button and select the object property from the pop-up menu and write the names of the F1, F2 and F3 factors appearing in Figure 25.

After that, by pressing error symbol to add residual to endogeneous variable. Finally, the plugins → Name Unobserved Variables menus are followed and the variable name is assigned to the error term of each indicator and the residue term of the endogeneous variable. Therefore, in the model shown in Figure 25, error names from e1 to e10 are automatically assigned by the program. Right click on the variables on the screen, and when the object properties is selected in the drop-down menu, the window shown in Figure

26 opens. In this window in text tab, variable names can be assigned.

In Figure 25, F1, F2, F3 and R1 variable names are assigned in this way. In this tab, the font size can be adjusted. In this window in the parameters tab, it is possible to assign the values of parameters which is desired to be kept constant in the model. F1 and F2 are exogeneous variables. Due to the fact that between them covariance is inserted by using the symbol . F3 is endogeneous variable. Therefore residual term (R1) is added by using the symbol . After the model is drawn, the Analysis Properties menu opens in order to select the analysis that you want to perform.

Click on the icon to open this menu. When this icon is clicked, the window shown in Figure 27 opens. The options marked on the Output tab in Figure 27 must be noted. On the Estimation tab, the Maximum likelihood option is usually selected. After the selections are made, the program is run by clicking on the icon. The special cases in which other estimation methods can be used are mentioned in previous chapters.

The items to be marked on the Output tab vary according to the properties of the analysis to be performed. For example, when performing confirmatory factor analysis, the factor score weights option should be marked. After running the analyzes, click the icon to view the results on the model. Click on the view text icon to view the results in tabular form.

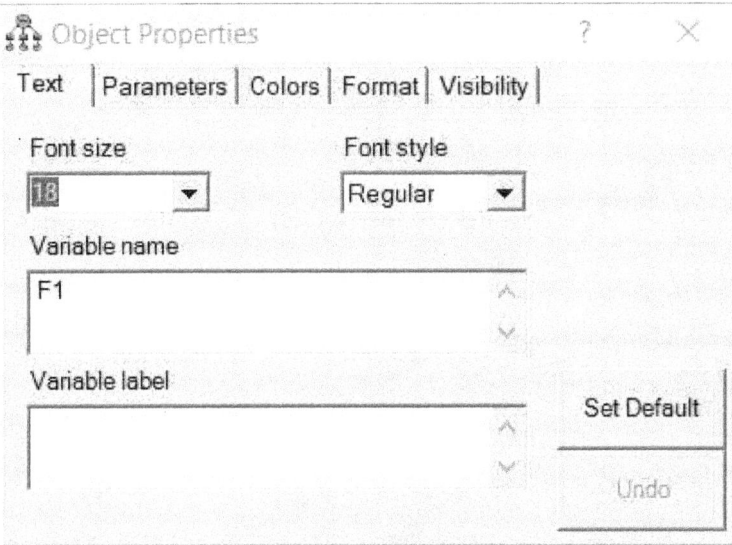

Figure 26. Object Properties Window Text Tab

When this icon is clicked, the screen shown in Figure 28 opens. For this window to be opened separately, the results can be viewed both in text format and on the model at the same time as shown in Figure 29. If you click on the Model Fit tab in the menu on the left side of this screen, the values of the fit indices of the model are reached. Regression weights are reached from the Estimate tab. Modification suggestions can be obtained from the modification indices tab.

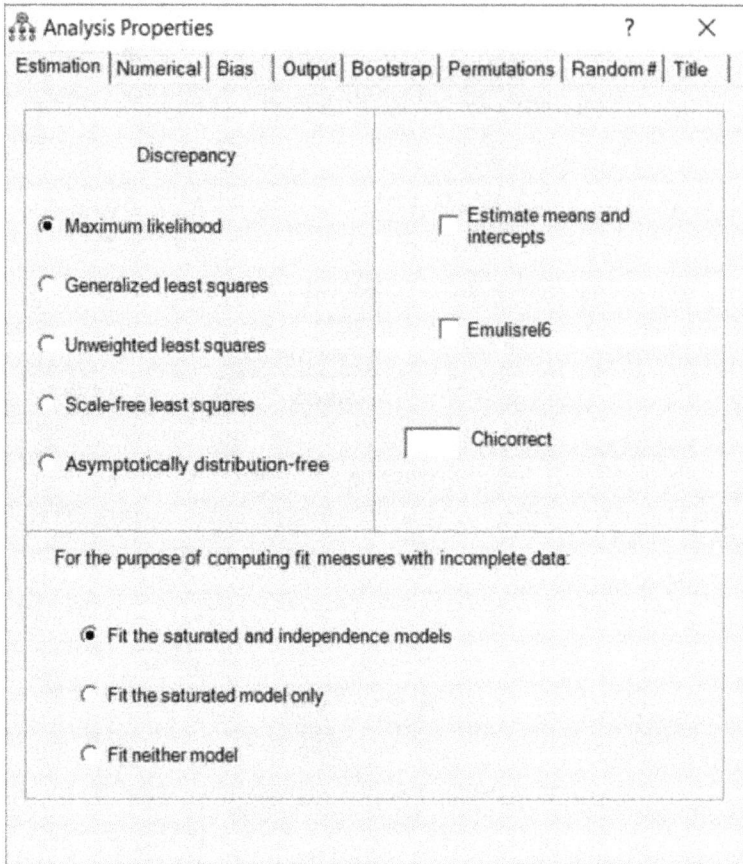

Figure 27. Analysis Properties Window Estimation Tab

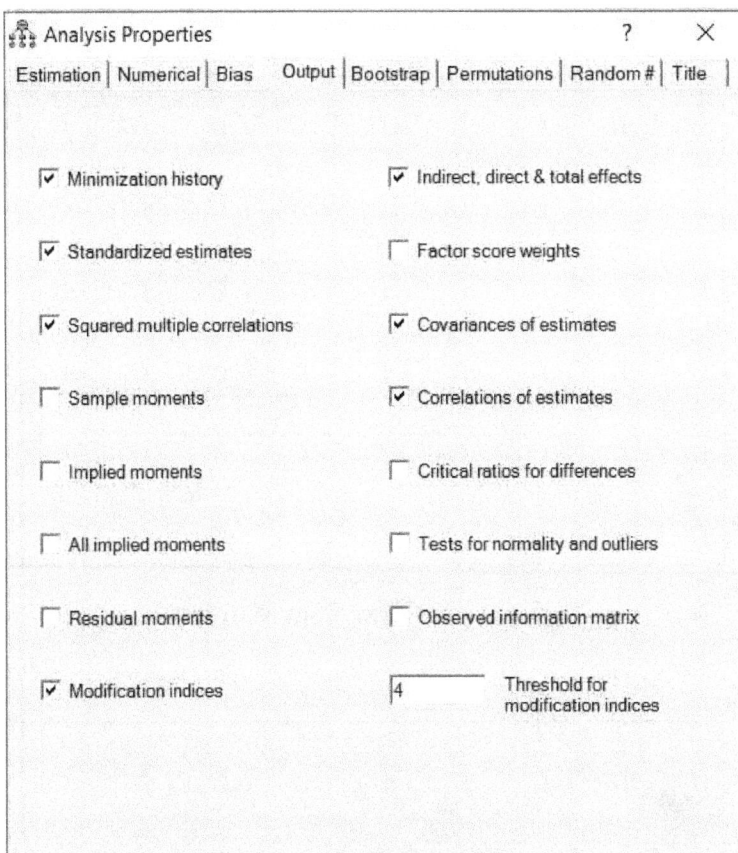

Figure 27. Analysis Properties Window Output Tab (Continued)

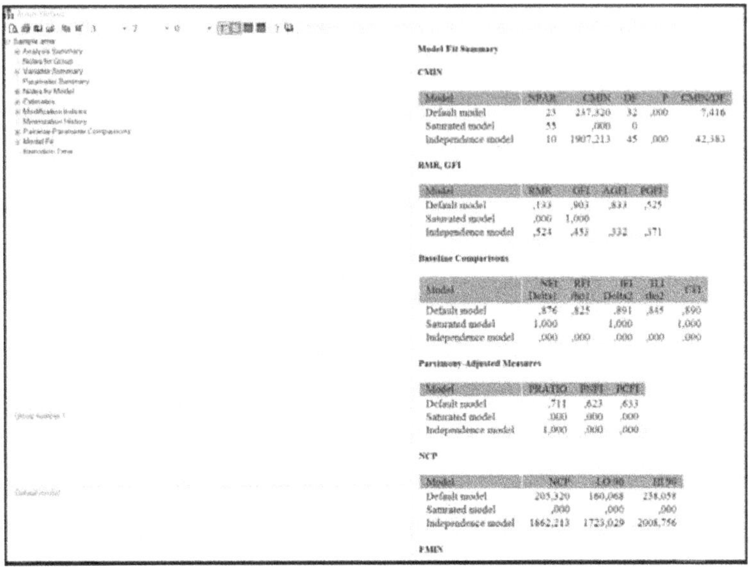

Figure 28. Text View Window

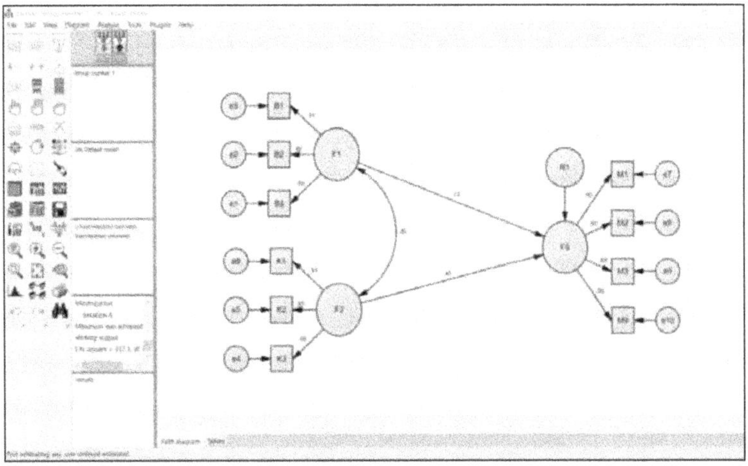

Figure 29. Showing Results on the Model

9. SAMPLE OF MEDIATOR VARIABLE ANALYSIS

In Section 7, we focused on the methodology of mediator variable analysis. Mediator analysis methodology is based on the hierarchical regression method introduced by Baron and Kenny in 1986 (Baron & Kenny, 1986). The structural equation modeling method allows analysis of direct and indirect relations together, so the application of the Baron and Kenny method in the structural equation model provides an advantage especially when there are more than one mediator variable in the conceptual model.

In Section 7, how the mediator analysis methodology is applied is described both by regression analysis method and by structural equation modeling method. In the case of mediator variable analysis in the structural equation modeling, a real research example is summarized below in order that the reader can better understand the way the tables are given (Civelek, İnce, & Karabulut, 2016). The literature section of the research is not given and only the parts enough to understand the method are given.

9.1. Title of the Research

Mediator role of attitude towards site in the relationship between system quality and net benefit.

9.2. Purpose

The aim of this research is to determine the role of users' attitudes towards the site and satisfaction levels in system quality which was found to have a positive impact on net benefit in previous research.

9.3. Conceptual Model and Scales

While constructing the conceptual model, the models developed in previous researches measuring the success of information systems were used.

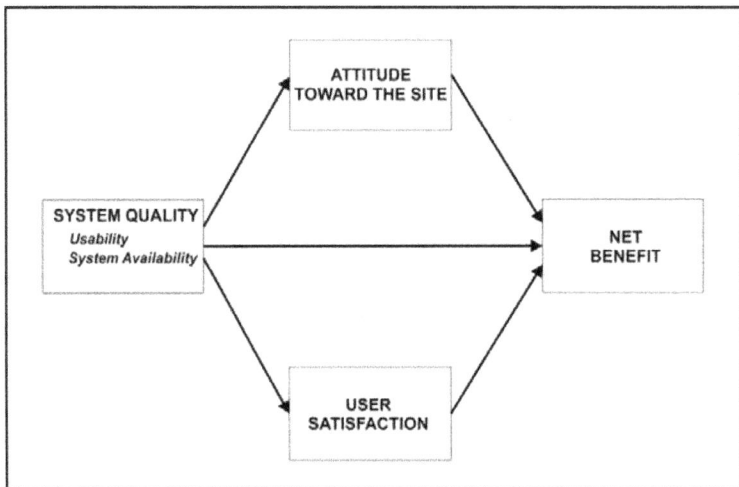

Figure 30. Conceptual Model

The scale developed by Wu and Wang in 2006 was used to measure the net benefit dimension (Wu & Wang, 2006). Scales developed by Chen et al. In 2013 were used to measure other dimensions (Chen, Rungruengsamrit, Rajkumar, & Yen, 2013). The scales were measured according to the five-point Likert scale. The conceptual model of the research is shown in Figure 30.

9.4. Determination of Validity and Reliability

Confirmatory factor analysis (CFA) was performed to determine the construct validity of the scales used. The fit indice values of the CFA model were found satisfactory: $\chi2/DF$ =3.790, CFI=0.920, IFI=0.921, RMSEA= 0.083. Table 10 shows the standard factor loadings of the confirmatory factor model. The standard factor loads are above 0,50 and the fit indices are close to the threshold values. Therefore the convergent validity of the scales is determined.

The AVE (Avarage Variance Extracted) values given in Table 12 are above the 0.50 threshold and the square root of the AVE values is greater than the correlation values in that column for each dimension. Therefore the discriminant validity of the scales is determined.

In addition, Cronbach α and Composite Reliability values are above the threshold value of 0.70, indicating the reliability of the scales used.

Table 11. Confirmatory Factor Analysis Results

Constructs	Items	Standardized Factor Loads	Unstandardized Factor Loads
Usability	Use1	0.760	1.00
	Use2	0.786	1.10
	Use3	0.647	0.80
System Quality	Sa1	0.912	1.00
	Sa2	0.807	0.91
	Sa3	0.663	0.62
Attitude Toward the Site	Ats1	0.717	1.00
	Ats2	0.747	1.13
	Ats3	0.847	1.11
	Ats4	0.763	1.10
User Satisfaction	Us1	0.761	1.00
	Us2	0.831	1.09
	Us2	0.941	1.25
	Us4	0.818	1.06
Net Benefit	Nf1	0.772	1.00
	Nf2	0.630	0.99
	Nf3	0.818	1.15
	Nf4	0.646	0.95

Note: For all $p<0.01$

9.5. Analysis Results

Three separate models have been analyzed as described in Chapter 7. The analysis results of three different models are compared in Table 13. In Figure 31, the results of the analysis of model 3 are given as an image.

Table 12. Correlation, AVE and Reliability Values

Variables	1	2	3	4
1. System Quality	(0.767)			
2. User Satisfaction	0.575*	(0.840)		
3. Attitude Toward Site	0.566*	0.677*	(0.770)	
4. Net Benefit	0.492*	0.672*	0.709*	(0.721)
Cronbach α	0.821	0.901	0.848	0.808
Composite Reliability (CR)	0.895	0.905	0.853	0.810
Avarage Variance Extracted (AVE)	0.589	0.706	0.593	0.520

*p < 0.01

Note: The values in brackets indicate the square root of the AVE values.

Table 13. Test Results

Relations	Model 1	Model 2	Model 3
System Quality →Net Benefit	0.701*		-0.03**
System Quality → User Satisfaction		0.733*	0.839*
System Quality → Attitude Toward Site		0.487*	0.880*
User Satisfaction →Net Benefit			0.331*
Attitude Toward Site → Net Benefit			0.651*
Model Fit Indices	χ2/df=4.039 CFI=0.974 IFI=0.974,RMS EA=0.08	χ2/df=3.678 CFI=0.967 IFI=0.967 RMSEA=0.08	χ2/df=3.750 CFI=0.944 IFI=0.944 RMSEA=0.08

Note: Regression coefficients are standard values.
*p<0.01
**Insignificant

As shown in Table 13, in the first model, the regression coefficient of the relationship between system quality and net benefit is statistically significant and quite high. However, in the third model, when the relationship includes user satisfaction and attitude toward site dimension, the coefficient of the relationship decreases and it turns out to be statistically insignificant.

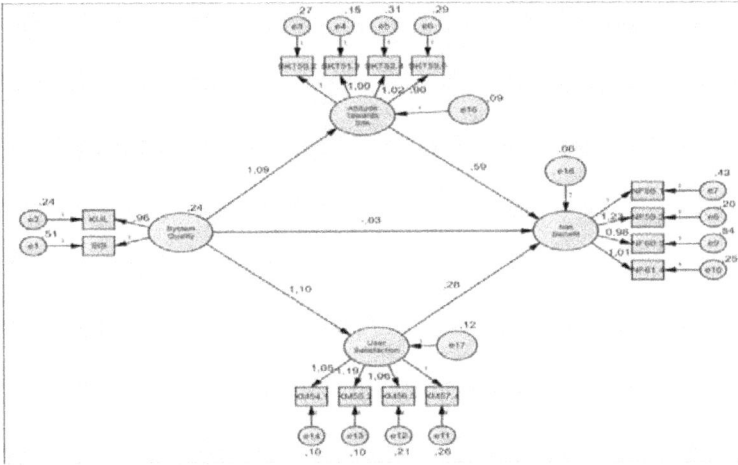

Note: χ2/DF = 3.750, CFI = 0.944, IFI = 0.944, RMSEA=
0.082

Figure 31. Model 3

9.6. Conclusion

As a result, this research shows that the quality of the system has no direct effect on the net benefit perceptions of the users. The system quality indirectly affects the net benefit perception. In other words, the relation between the system quality and the net benefit perception user satisfaction and attitudes of the users towards the site play mediator role. This means: Increasing the quality of the system primarily increases the satisfaction of users and improves their attitude towards the site. In the end, user satisfaction and attitude, which turned into positive, increase the net benefit perception.

10. MULTITRAIT-MULTIMETHOD MODEL

In some cases, nested structures can be found. In other words, some scale questions can be designed to measure nested concepts at the same time. As an example, assume that a scale to measure teachers' competence in a school has been developed. On this scale, there are questions that measure the social, academic and English competence of the teachers. The questions were asked to the same number of groups of students and colleagues. In this example, there is a structure consisting of three features and two separate methods as shown in Figure 32, the best method that can be used to determine the construct validity of such models is Multitrait-Multimethod Model. As shown in Figure 32, there are two separate questions asked to two different groups. In the model shown in Figure 32, social, academic and English competences are the traits and students and peers are the methods.

In the model there are as many questions as the multiplication of the numbers of traits and methods. Multitrait-Multimethod Model was first proposed by Campbell and Fiske in a paper published in 1959.

Although different alternatives have been proposed over time models based on covariance structure became important (Byrne, 2010).

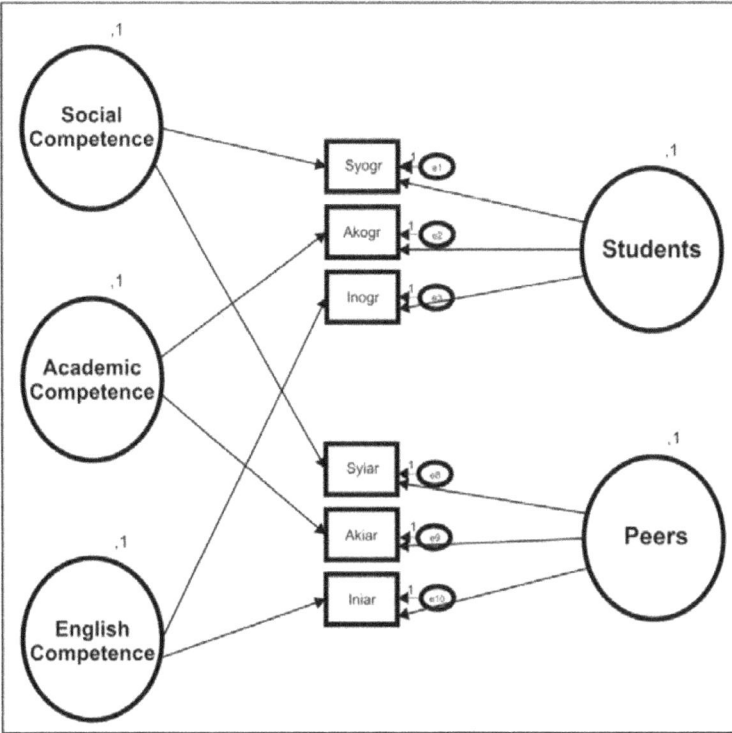

Figure 32. Multitrait-Multimethod Model

Source: Byrne, B. M. (2010). *Structural Equation Modeling with AMOS. New York: Routledge Taylor & Francis Group.*

Construct validity, according to Campbell and Fiske, focused on the determination of the convergent validity, which expresses the correlation of the components that make up, the discriminant validity, which expresses low degree of correlation with the components constituting other constructs and method effect, which is an extension of discriminant validity (Campbell & Fiske, 1959). The method effect can be defined as the bias that results from using the same method to evaluate different properties (Byrne, 2010).

In Figure 33, there is a larger example that will provide a better understanding of Multitrait-Multimethod Model. In the figure, there is an assumption model of this example. In Figure 33, there are 7 traits as X, Y, V, W, Z, T, U and 5 methods as A, B, C, D, E. Therefore, a scale consisting of 35 questions was used. The notation of ",1" on the latent variables forming traits and methods indicates that the factor variance is fixed at 1 (The another notation "1," indicates that the factor averages are fixed at 1). This is a symbolic notation, and comma notation does not appear in the AMOS program, it looks as seen in Figure 35. To fix the variance of a hidden variable in the AMOS program, enter the value in the variance box under the "Parameter" tab in the "Object Properties" dialog box which opened by right mouse button on selected variable. When we look at the parameter summaries in Table 14, it is seen that the variances of 12 variables are kept constant.

It is also seen that the variance of 35 error terms obtained as a result of estimation is released. 70 regression coefficients were estimated and 35 regression coefficients belongs to the error terms were fixed. Therefore, there are a total of 105 regression weights. In this case, when the fixed regression weights and latent variable variances are evaluated together, it is seen that 47 parameters are kept constant. There are totally 183 parameters in the whole model.

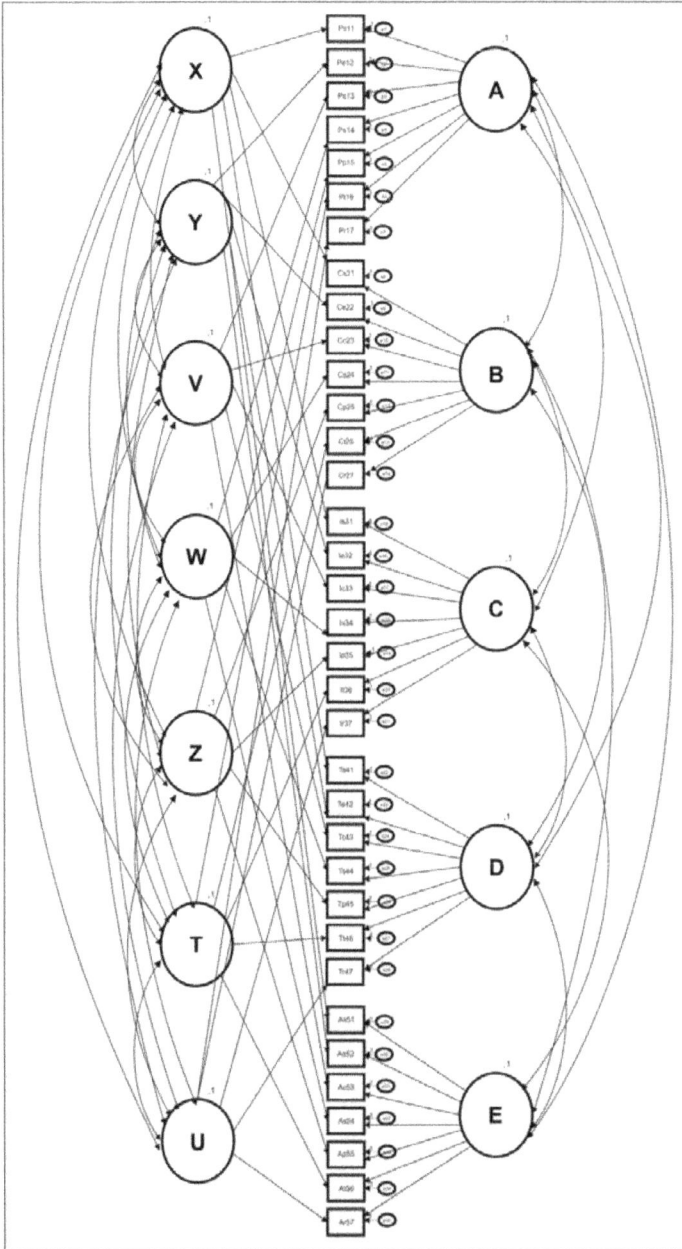

Figure 33. Hypothesized Multitrait-Multimethod Model

Table 14. Parameter Summary

	Weights	Covariances	Variances	Total
Fixed	35	0	12	47
Labeled	0	0	0	0
Unlabeled	70	31	35	136
Total	105	31	47	183

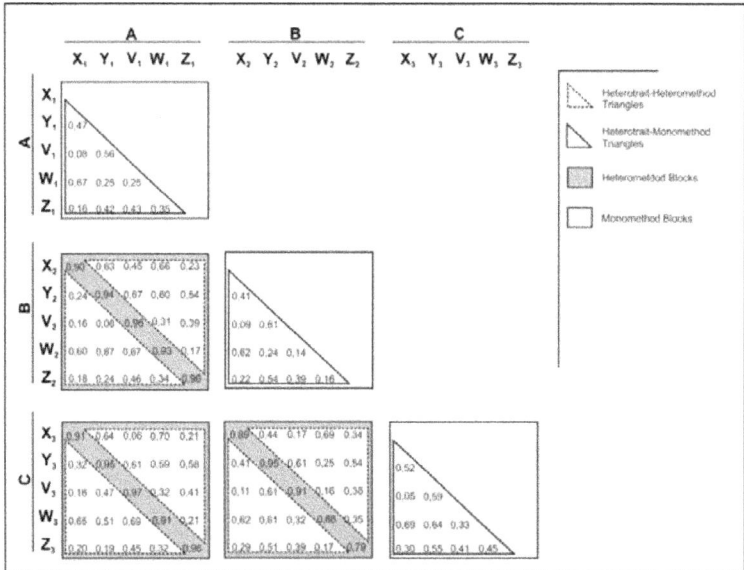

Figure 34. Multitrait-Multimethod Matrix

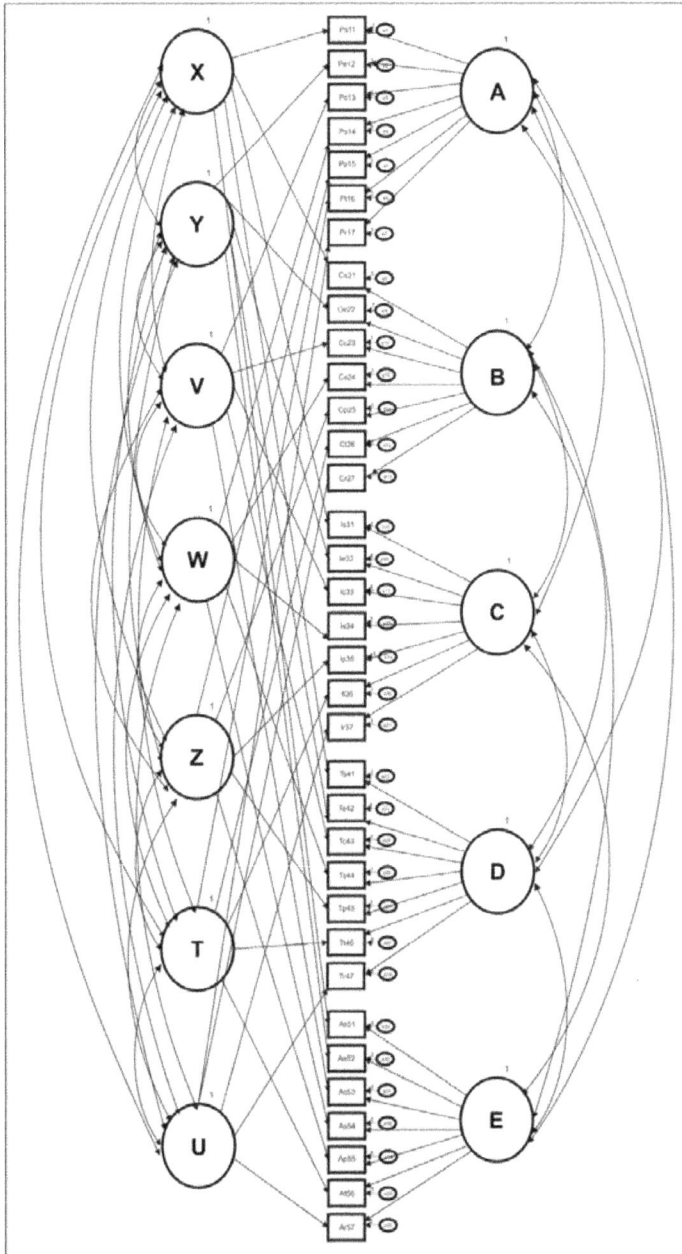

Figure 35. Model 1 (freely correlated traits; freely correlated methods)

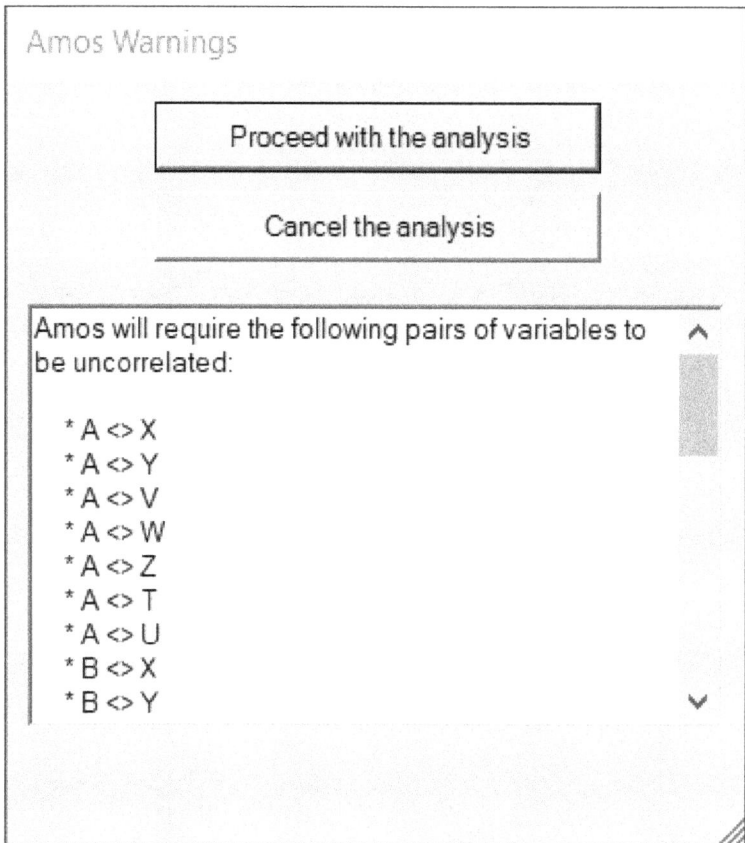

Figure 36. AMOS Warnings Window

The matrix structure in the background of the example model in Figure 33 is as seen in Figure34. Due to the difficulty in displaying the figure, only a limited drawing including methods A, B and C was made.

Table 15. Summary of Fit Indices

Fit Indices

Models	x^2	df	CFI	RMSEA	90% C.I.	PCLOSE
1. Freely correlated traits; freely correlated methods	86.62	78	.897	.015	.000 .048	.897
2. No traits; freely correlated methods	459.12	98	.693	.204	.122 .157	.000
3. Perfectly correlated traits; freely correlated methods	317.12	85	.795	.086	.081 .110	.000
4. Freely correlated traits; uncorrelated methods	123.39	81	.964	.058	.037 .065	.000

In figure1 34, Heterotrait-Heteromethod Triangles, Heterotrait-Monomethod Triangles, Heterometdod Blocks and Monomethod Blocks in Multitrait-Multimethod Matrix are shown. When the Multitrait-Multimethod models are run, the warning shown in Figure 36 appears. The reason for this warning is that there are no covariances among all the exogeneous variables. In this case, "proceed with the analysis" should be selected and continued.

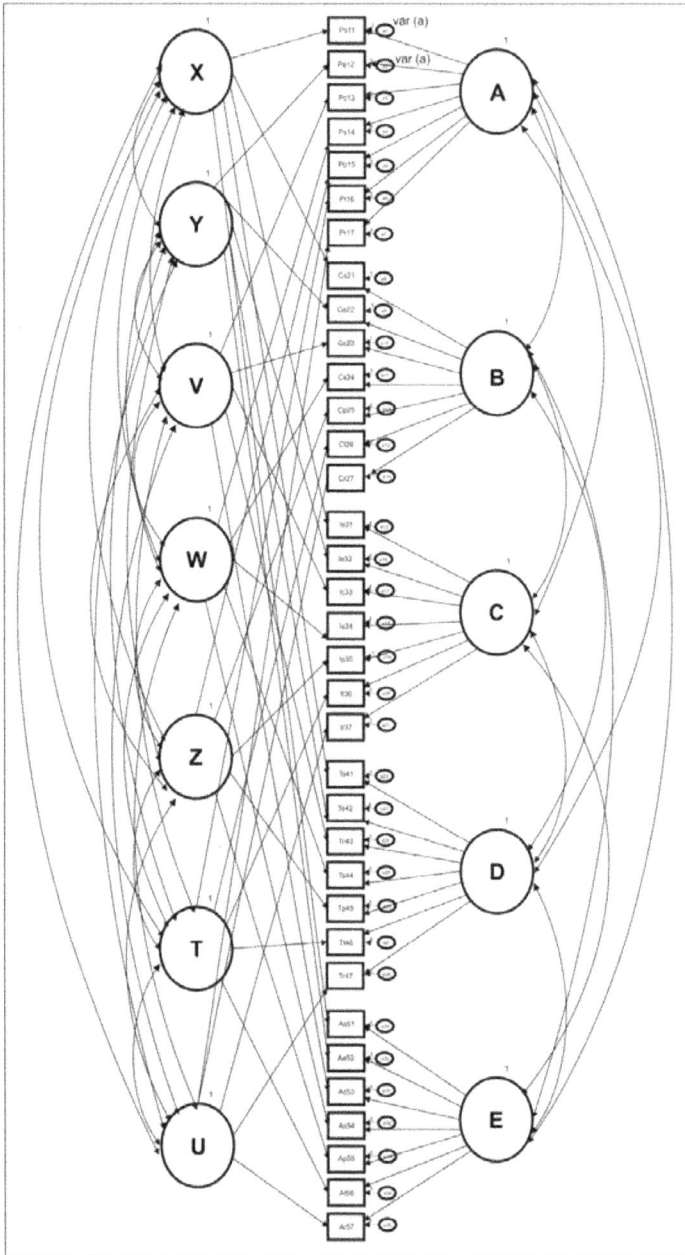

Figure 37. Post Hoc Model 1

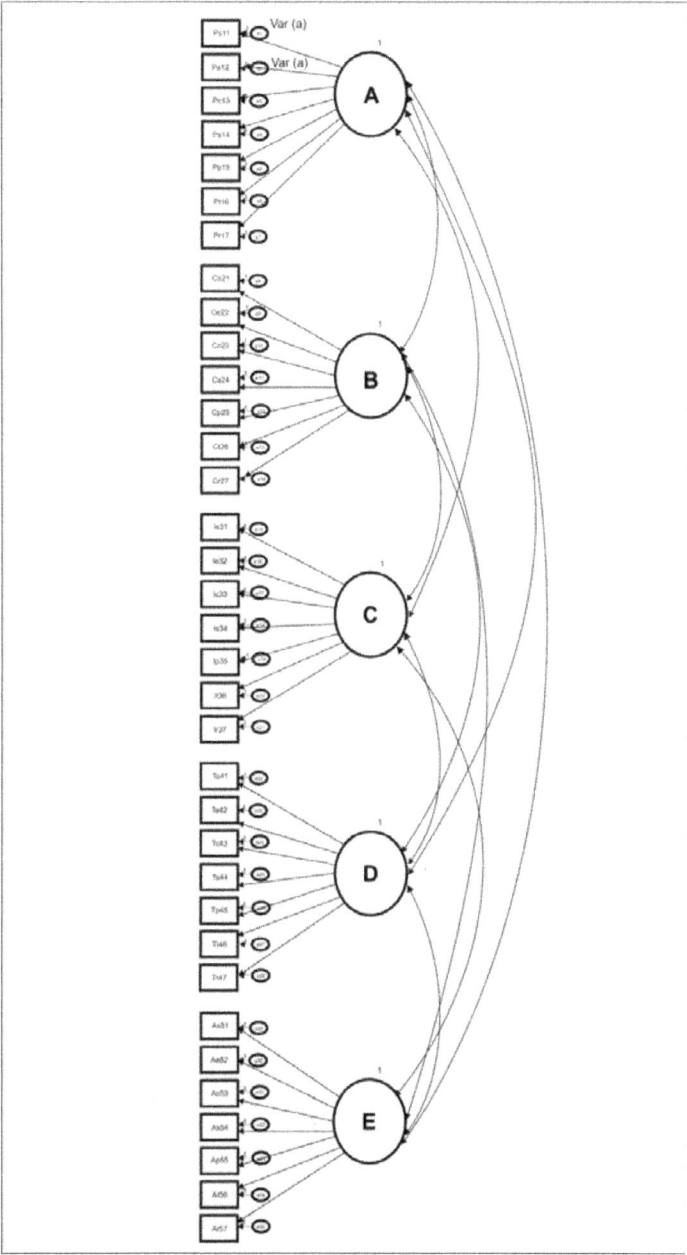

Figure 38. Model 2 (no traits; freely correlated methods)

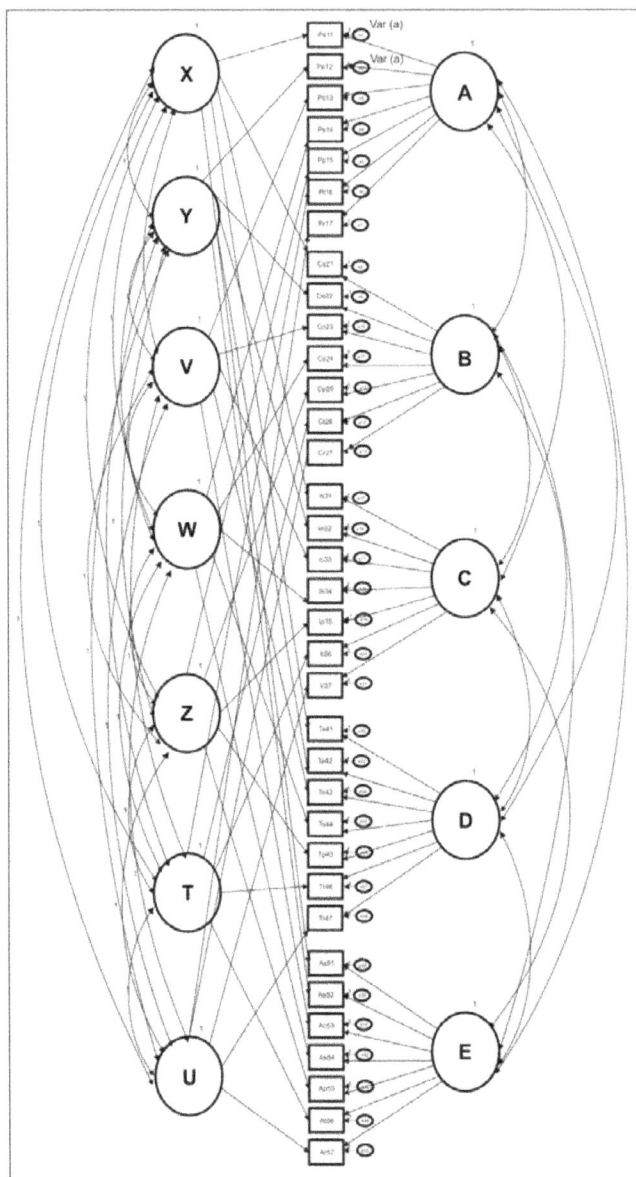

Figure 39. Model 3 (perfectly correlated traits; freely
correlated methods)

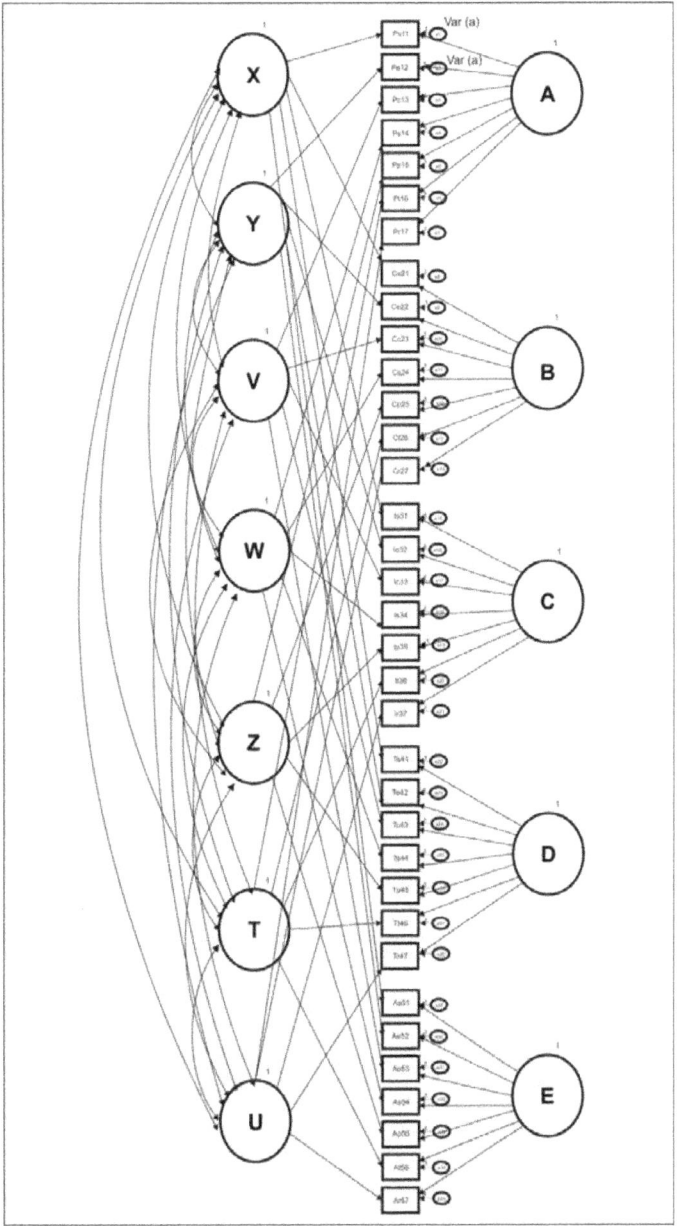

Figure 40. Model 4 (freely correlated traits; uncorrelated methods)

In some cases, the error terms get negative values when the model is run. In this case, a post hoc model is created as shown in Figure 37.

In this model, Var (a) value is assigned in order to fix the negative error term. The assignment of value Var(a) is also made in alternative models created for comparison with Model 1. Note that the same value assignment is shown in Figure 38.-39.-40.

Table 16. Comparisons of Nested Models

Models	Difference		
	x^2	df	CFI
Test of Convergent Validity			
Model 1 and Model 2 (traits)	372.506	20	.204
Test Discriminant Validity			
Model 1 and Model 3 (traits)	230.502	7	.102
Model 1 and Model 4 (methods)	74.230	3	.067

Then, 3 alternative models are created. These models are as seen in Figures 38-39-40. Model 2 consists only of methods. The latent variables that make up the traits are not included in the model. The parameters of the covariances between the methods are released. In Model 3, the features and methods are included in the same model but this time the covariance parameters between the latent variables forming the traits are fixed to 1.

Table 17. Model 1 Trait and Method Loadings

	X	Y	V	W	Z	T	U	A	B	C	D	E
				A								
X	.920							.008				
Y		.901						.601				
V			.898					.007				
W				.794				.405				
Z					.854			.506				
T						.426		.522				
U							.324	.714				
				B								
X	.401								.302			
Y		.306							.852			
V			.384						.701			
W				.399					.628			
Z					.424				.701			
T						.789			.574			
U							.698		.358			
				C								
X	.640									.406		
Y		.501								.720		

	1	2	3	4	5	6	7	
V		.654						.506
W			.701					.525
Z				.689				.603
T								.745
U						.754		.289
							.597	.374

D

	1	2	3	4	5	6	7	
X	.256							.356
Y		.406						.902
V			.250					.604
W				.274				.712
Z					.372			.586
T						.755		.457
U							.743	.771

E

	1	2	3	4	5	6	7	
X	.595							.411
Y		.489						.398
V			.525					.549
W				.424				.375
Z					.445			.601
T						.379		.832
U							.408	.566

Note: Estimates are standardized

For this reason this model is called as perfectly correlated traits; freely correlated methods. In Model 4, there is no covariance between methods. In Model 1, which is the model where the correlations between traits and methods are free, calculated correlation values are shown in Table 18. In the Multitrait-Multimethod models approach, analyzes are performed at the matrix level and the parameter level to test the construct validity (both convergent and discriminant validity).

In matrix level analysis, comparison of fit indices of alternative models with model 1 is made. Table 15 shows the values of fit indices of each model. Table 16 compares each alternative model with model 1. The comparison between Model 1 and Model 2 shows convergent validity, comparisons between Model 1 and Model 3 and Model 4 show the discriminant validity. The significance of difference between X^2 values of Model 1 and Model 2 indicating the convergent validity is sufficient.

As shown in Table 16, ΔX^2 (372.506, p<0.01) and ΔCFI (0.204, p<0.01) are significant. In addition, the difference between X^2 values of Model 1 and Model 3 and Model 4 are the basis for confirming the discriminant validity. As shown in Table 16, between Model 1 and Model 3, ΔX^2 (230.502, p<0.01) and ΔCFI (0.102, p<0.01) are significant, between Model 1 and Model 4, ΔX^2 (74.230, p<0.01) and ΔCFI (0.067, p<0.01) are significant. Another indicator of construct validity is parameter level analysis. For comparison at the parameter level, the factor loadings and factor correlations in Table 17 are compared.

Table 18. Traits and Methods Correlations in Model 1

Construct	Traits							Methods				
	X	Y	V	W	Z	T	U	A	B	C	D	E
X	1											
Y	.345	1										
V	.302	.789	1									
W	.220	.720	.487	1								
Z	.351	.698	.501	.607	1							
T	.455	.521	.201	.421	.584	1						
U	.248	.836	.478	.225	.160	.370	1					
A								1				
B								.197	1			
C								.201	.421	1		
D								.254	.428	.222	1	
E								.218	.648	.168	.334	1

Factor loads are expected to be significant. the Multitrait-Multimethod models, four models are formed as described above. There is a need for a model that is free from bias in these models. In this case, correlated uniqueness model can be established. In Figure 41, correlated uniqueness model is shown.

First, it is checked whether the fit indices of the model are within the acceptable range. In this model, the factor loads must be significant in order to confirm the construct validity.

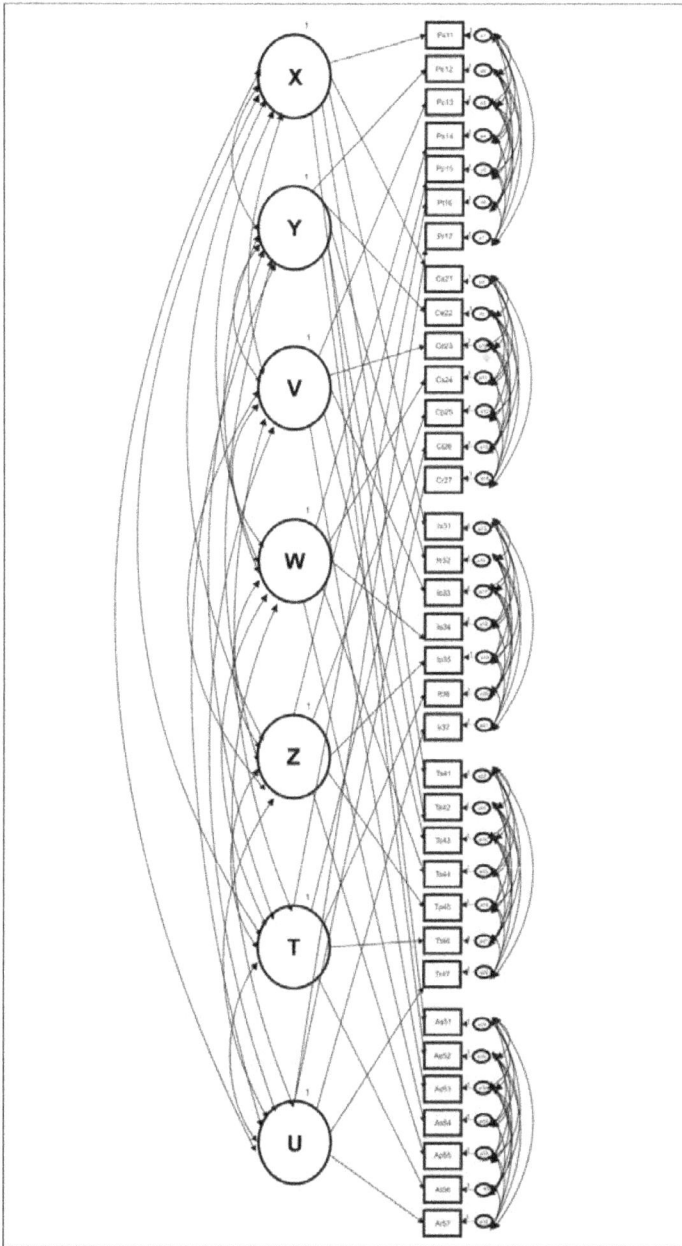

Figure 41. Model 5 (Correlated Uniqueness Model)

Table 19. Factor Loading in Model 5

	X	Y	V	W	Z	T	U
				A			
X	.810						
Y		.801					
V			.797				
W				.644			
Z					.751		
T						.800	
U							.721
				B			
X	.501						
Y		.416					
V			.353				
W				.489			
Z					.518		
T						.468	
U							.499
				C			
X	.741						
Y		.428					
V			.555				
W				.721			

Z					.489		
T						.706	
U							.742

D

X	.368						
Y		.517					
V			.398				
W				.277			
Z					.365		
T						.479	
U							.398

E

X	.498						
Y		.489					
V			.520				
W				.328			
Z					.531		
T						.544	
U							.298

Note: Estimates are standardized

11. METHODS TO BE APPLIED IN CASE OF DATA INADEQUACY

Sometimes, there may be cases where the assumptions of the estimation methods used are not met by the existing data set. In this case, there are methods that can be applied if it is necessary to be satisfied with the dataset available. Leading methods among them are bootstrap partial least sequare.

11.1. Bootstrap Method

The bootstrap technique is applied when one of the assuptions of normal distribution or being continuous variable is not met. This method was developed by B. Efron in 1979 (Efron, 1979). In many studies in the literature the condition of normal distribution obligation is neglected. It is also seen in many studies in the literature that X^2 value is derived by maximum likelihood and generalized least squares methods. Estimation methods which are frequently used in the structural equation model are these two methods. In particular, with the non-normal distribution, the number of observations is also low cause X^2 value to increase. At the same time, irreversible and inadequate modifications made during the analysis of such data are not scientifically acceptable and result in inconsistent estimations about the population. In the bootstrap method, a different data set is obtained from the existing observations (Sacchi, 1998). This method is basically the derivation of the sample from the sample.

There are advantages and limitations of the bootstrap process. The main advantage of the bootstrap technique is the ability to evaluate the accuracy of the predicted parameters. The idea underlying the bootstrap technique is to create sub-samples of the current data and look at the distribution of the parameters computed from each sub-sample.

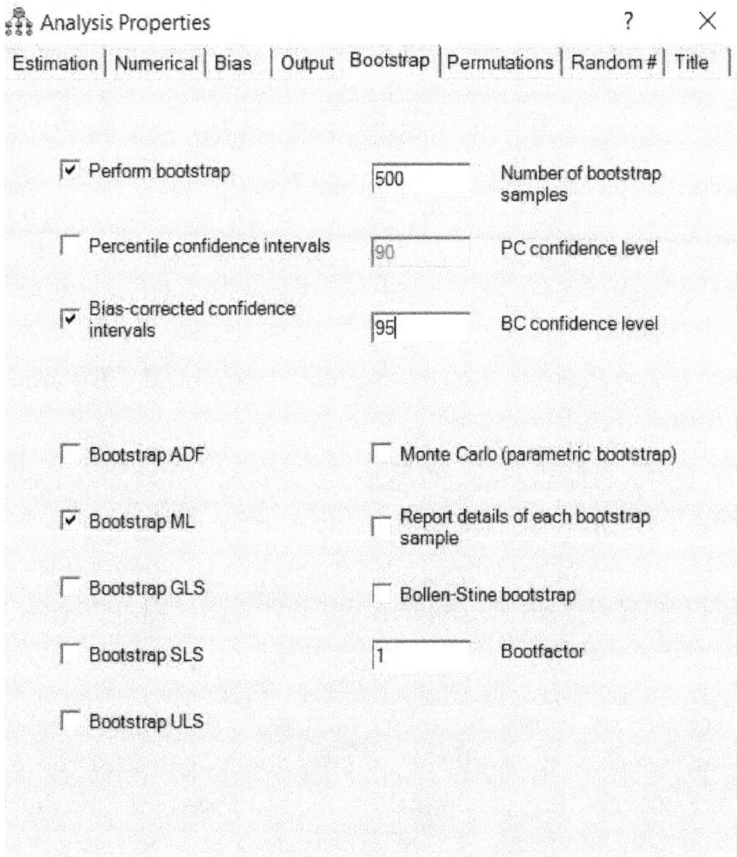

Figure 42. Bootstrap Properties Window

There are various criticisms in the literature about the correctness of the results obtained with this technique. (Kline, 2011) (Ichikawa, 1995) (Yung & Bentler, 1994) (Hancock & Nevitt, 1999). It's wrong to see this technique as a magical method. This method should not be used especially in small samples with low representation ability and in extreme non-normal distribution (Kline, 2011). Because especially in small samples, there is the posibility to further enhance its properties that do not match the population (Rodgers, 1999). The AMOS program includes bootstrap analysis. Figure 42 shows the bootstrap tab under the analysis properties window. In this window it is initially marked that how much the bootstrap technique can be applied in the sample. The window is closed after the confidence interval and prediction method are selected.

11.2. Partial Least Square Structural Equation Modeling (PLS-SEM)

It is also called covariance-based structural equation modeling since the structural equation model that has been examined in the previous sections is based on the covariance matrix. However partial least square structural equation modeling is based on variance. For this reason, it is also called as the variance-based structural equation modeling. Partial least square structural equation modeling (PLS-SEM) is an advantageous method when the assumptions of least squares are not met. It is an alternative of covariance-based structural equation modeling (CB-SEM). It is a second generation multivariate analysis method that enables measurement model and structural model to be analyzed together like covariance-based structural equation modeling.

Table 20. Classification of Multivariate Methods

	Exploratory	Confirmatory
First Generation Techniques	• Cluster Analysis • Exploratory Factor Analysis • Multidimensional Scaling	• Analysis of Variance • Logistic Regression • Multiple Regression • Confirmatory Factor Analysis
Second Generation Techniques	• Partial Least Square Structural Equation Modeling (PLS-SEM)	• Covariance-Based Structural Equation Modeling (CB-SEM)

Source: Hair, J., Hult, G., Ringle, C., & Sarstedt, M. (2017). *A primer on partial least squares structural equation modeling PLS-SEM. Los Angeles: SAGE.*

But it is not a confirmatory analysis technique like Covariance-Based Structural Equation Modeling. Table 20 shows the classification of multivariate methods (Hair, Hult, Ringle, & Sarstedt, 2017). As shown in Table 20, it is an exploratory analysis technic.

According to some sources in the literature, covariance-based structural equation modeling is a more powerful and reliable method. For this reason, the partial least square structural equation modeling method is generally preferred in cases where the conditions listed below are found:

• If the sample is small.

• If the data do not distribute normally.

• If the number of indicators connected to the latent variable is less than three.

• If there is a multicollinearity.

• There is missing value.

• If the number of observations is less than the number of explanatory variables.

If the above listed conditions are found, method PLS-SEM method is far superior to method CB-SEM. Because, in these cases, it reduces the unexplained variance to the lowest level. As the model is complex, such as in the CB-SEM method, no larger sampling is required in PLS-SEM. However, some researchers who have done research on the sampling sensitivity of the PLS-SEM method have raised the ten-fold rule. According to this rule, there is a necessity to have 10 times observation of the number of indicators used to measure a construct in the measurement model and 10 times observation the number of the path in a structural model (Barclay, Higgins, & Thompson, 1995).

However, the PLS-SEM method is a non-parametric method because it does not have any distributional assumption (Hair, Hult, Ringle, & Sarstedt, 2017). It is also an explanatory approach, which is why it is preferred in exploratory research. In other words, when the theory is underdeveloped, it can be said that researchers prefer to use partial least squares structural equation modeling. This judgment is partially correct in cases where the structure need to be predicted and relations need to be explained (Rigdon, 2012).

When the theory needs to be tested and verified, in case of there is cycles in the structural model and if the model needs to be verified in general with fit indices it is more accurate to use CB-SEM method. Because the PLS-SEM method can not explain loop-related relations. In addition, it does not give general fit indices of the model.

Partial least squares method can be easily implemented by means of a packet program called SmartPLS. SmartPLS is a packet program that allows the creation of partial least squares based structural equation models. Structural equation modeling programs outside of SmartPLS makes the maximum likelihood estimation method the default choice. Because covariance based structural equation modeling is defined by this estimation method in the literature. However, as explained in previous chapters, different estimation methods can be used in case of necessity. Therefore, the method that is most likely to be used in the covariance-based structural equation modeling should be based on a valid reason (Kline, 2011). This is the most important reason for criticism of PLS-SEM method. For this reason, it is necessary to make sure that all solution alternatives are exhausted if the partial least squares method is used in a research according to some sources. But there are also resources that assess them as blind and misleading criticisms, and that reveal the advantages of the partial least squares method (Henseler, Dijkstra, Sarstedt, Ringle, Diamantopoulos, & Straub, 2014). Despite all this criticism and hesitation, the PLS-SEM method has become an increasingly used method in scientific studies (Hair, Hult, Ringle, & Sarstedt, 2017).

As mentioned above, a packet program called SmartPLS is used to construct the least squares based structural equation models. The following steps describe how to build a structural equation model with SmartPLS. These descriptions are based on SmartPLS 3 version. When SmartPLS is first turned on, the screen shown in Figure 43 'opens. In order to create a new project, in the top menu "File" is selected, and after that "Create New Project" command is clicked on.

Figure 43. SmartPLS Start Screen

When the command is clicked, the "Create Project" window shown in Figure 44 opens. Enter the project name in the Name field of this window and press the OK button to close the window. As a result, the project name is displayed in the Project Explorer section of the main screen. Double click on "Double-click to import data" on the project name and the data set to be worked on is connected to the project. The SmartPLS program is not a SPSS-compatible program like AMOS, so it only accepts Excel files. If the data set to be used is in the SPSS program, data can be easily transferred from the SPSS to the Excel file.

Figure 44. Project Creation Screen

After the data set has been connected, in the indicator section, indicators are displayed as listed. Once you click on the project name in the Project Explorer section, the white screen opens again. When the desired indications in this segment are selected in groups and are dragged in to the middle of screen, it can be seen that the latent variable and its indicators are automatically drawn as shown in Figure 45.

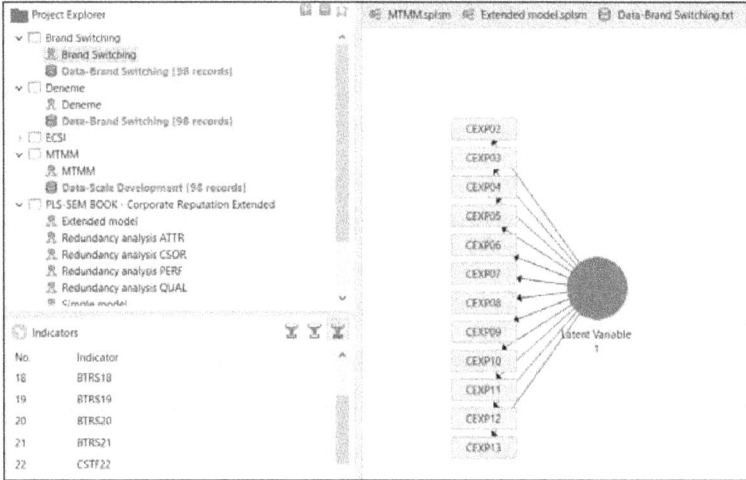

Figure 45. Drawing Latent Variable

The hidden variable plotted in Figure 45 can be easily dragged with the mouse and the variable can be renamed by right clicking on it. After all the variables are drawn in this way, paths are drawn between the hidden variables using the Connect button at the top of the screen. After the model drawing process is completed, the Calculate → PLS Algorithm command, which is located at the top of the screen, is executed. The predicted values of the path coefficients in the opened window can be obtained in matrix form or graphically. Figure 46 shows the estimated values in a matrix form.

Path Coefficients

Matrix | Path Coefficients

	Attitude_to ...	Brand_Trust	Customer _...	Customer_L...	Customer_S...	Intention to...
Attitude_to ...						0.457
Brand_Trust				0.254		
Customer _E...		0.447		0.189	0.442	
Customer_L...	-0.327					-0.250
Customer_S...				0.303		
Intention to ...						

Figure 46. Coefficient Estimation Results

If you click on the project name again in the Project Explorer pane, all the coefficient values can be displayed on the main model, including the measurement model, as shown in Figure 47. In this Figure, the values in the middle of each latent variable indicate R^2 values. As a result, analysis reports can be written by interpreting these values.

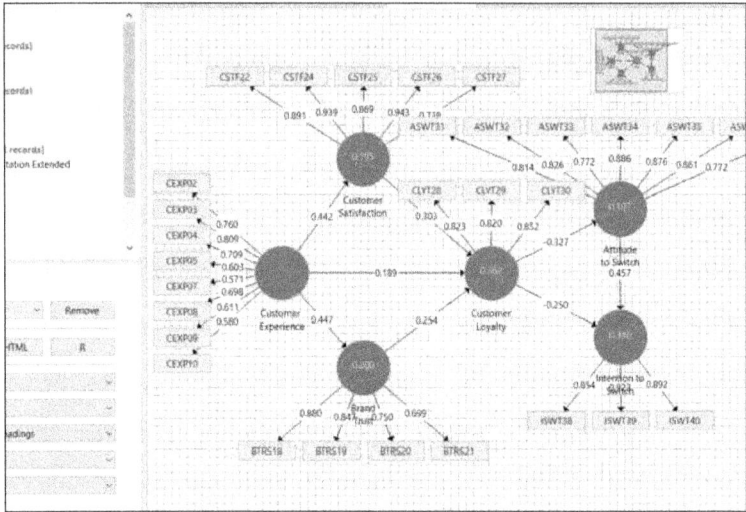

Figure 47. Estimation Results

110

BIBLIOGRAPHY

Aksu, G., Eser, M. T., & Güzeller, C. O. (2017). *Açımlayıcı ve Doğrulayıcı Faktör Analizi ile Yapısal Eşitlik Modeli Uygulamaları.* Ankara: Detay Yayıncılık.

Anderson, J., & Gerbing, D. (1988). Structural Equation Modelling in Practice: A Review and Recommended Two-Step Approach. *Psychological Bulletin.*

Avcılar, M., & Varinli, İ. (2013). *Perakende Marka Değerinin Ölçümü ve Yapısal Eşitlik Modeli Uygulaması* (1 b.). Ankara: Detay Yayıncılık.

Bagozzi, R. P., & Yi, Y. (1990). Assessing Method Variance in Multitrait-Multimethod Matrices: The Case of Self-reported Affect and Perceptions at Work. *Journal of Applied Psychology, 75*(1), 547-560.

Baltes, P., & Nesselroade, J. (1979). *History and rationale of longitudinal research.* New York: Academic Press.

Barclay, D., Higgins, C., & Thompson, R. (1995). The partial least sequares approach to causal modeling: Personal computer adoption and use as illustration. *Technology Studies, 2,* 285-309.

Baron, R., & Kenny, D. (1986). The Moderator-Mediator Variable Distinction in Social Phychological Research: Conceptual, Strategic and Statistical Considerations. *Journal of Personality and Social Phychology, 6*(51), 1173-1182.

Bayram, N. (2013). *Yapısal Eşitlik Modellemesine Giriş*. Bursa: Ezgi Kitapevi.

Bentler, P. M., & Chou, C.-P. (1987). Practical Issues in Structural Modeling. *Sociological Methods Research, 16*(1), 78-117.

Byrne, B. M. (2010). *Structural Equation Modeling with AMOS*. New York: Routledge Taylor & Francis Group.

Campbell, D., & Fiske, D. (1959). Covergent and Discriminant Validation by the Multitrait-Multimethod Matrix. *Psychological Bulletin, 56*, 81-105.

Chen, J., Rungruengsamrit, D., Rajkumar, T., & Yen, D. (2013). Success of Electronic Web Sites: A Comparative Study in Two Countries. *Information & Management, 50*(6), 344-355.

Civelek, M., İnce, H., & Karabulut, T. (2016). The Mediator Roles Of Attitude Toward The Web Site And User Satisfaction On The Effect Of System Quality On Net Benefit: A Structural Equation Model On Web Site Success. *European Scientific Journal*, 61-73.

Çelik, H. E., & Yılmaz, V. (2013). *Lisrel 9.1 ile Yapısal Eşitlik Modellemesi*. Ankara: Anı Yayıncılık.

Doğan, İ. (2015). *Farkli Veri Yapisi ve Örneklem Büyüklüklerinde Yapisal Eşitlik Modellerinin Geçerliği ve Güvenirliğinin Değerlendirilmesi*. Eskişehir: Eskİşehİr Osmangazİ Ünİversİtesİ Doktora Tezi.

Dursun, Y., & Kocagöz, E. (2010). Yapısal Eşitlik Modellemesi ve Regresyon: Karşılaştırmalı Bir Analiz. *Erciyes Üniversitesi İ.İ.B.F. Dergisi*(35), 1-17.

Efron, B. (1979). Bootstrap method: Another look at the jackknife. *Annals of Statistics, 7*, 1-26.

Fornell, C., & Larcker, D. (1981). Evaluating Structural Equation Models with Unobservable Variables and Measurement Error. *Journal of Marketing Research, 18*(1), 39-50.

Gerbing, D., & Anderson, J. (1988). An Updated Paradigm for Scale Development Incorporating Unidimensionality and Its. *Joournal of Marketing Research, 25*(2), 186-192.

Gujarati, D. (1999). *Essentials of Econometrics* (2 b.). Singapore: McGRAW-HILL.

Hair, J., Hult, G., Ringle, C., & Sarstedt, M. (2017). *A primer on partial least squares structural equation modeling PLS-SEM.* Los Angeles: SAGE.

Hancock, G., & Nevitt, J. (1999). Bootstrapping and the identifcation of exogenous latent variables within structural equation models. *Structural Equation Modeling, 6*, 394-399.

Henseler, J., Dijkstra, T., Sarstedt, M., Ringle, C., Diamantopoulos, A., & Straub, D. (2014). Common beliefs and reality about partial least squares: Comments on Rönkkö & Evermann (2013). *Organizational Research Methods, 17*, 182-209.

Ichikawa, M. (1995). Application of the bootstrap methods in factor analysis. *Psychometrika, 60*, 77-93.

Jayaram, J., Kannan, V., & Tan, K. (2004). Influence of initiators on supply chain value creation. *International Journal of Production Research, 42*(20), 4377-4399.

Karagöz, Y. (2016). *SPSS ve AMOS 23 Uygulamalı İstatistiksel Analizler.* Ankara: Nobel.

Kline, R. (2011). *Principles and practice of structural equation modeling* (3nd Ed. b.). New York: Guilford.

MacKinnon, D., Lockwood, C., Hoffman, J., West, S., & Sheets, V. (2002). A comparison of methods to test mediation and other intervening variable effects. *Psychological Methods, 7*, 83-104.

Meydan, C. H., & Şen, H. (2011). *Yapısal Eşitlik Modellemesi AMOS Uygulamaları.* Ankara: Detay Yayıncılık.

Raykov, T. (1997). Estimation of composite reliability for congeneric measures. *Applied Psychological Measurement, 21*(2), 173-184.

Raykov, T., & Marcoulides, G. (2006). *A First Course in Structural Equation Modeling.* Mahwah: Lawrence Erlbaum Associates.

Rigdon, E. (2012). Rethinking partial least squares path modelling: In praise of simple methods. *Long Range Planning, 45*, 341-358.

Rodgers, J. (1999). The bootstrap, the jackknife, and the randomization test: A sampling taxonomy. *Multivariate Behavioral Research, 34*, 441-456.

Sacchi, M. (1998). A bootstrap procedure for high-resolution velocity analysis. *Geophysics, 63*(5).

Sarstedt, M., & Mooi, E. (2014). *A concise guide to market research: The process, data, and methods using IBM SPSS statistics* (2nd b.). Berlin: Springer.

Schermelleh-Engel, K., Moosbrugger, H., & Müller, H. (2003). Evaluating the Fit of Structural Equation Models: Tests of Significance and Descriptive Goodness-of-Fit Measures. *Methods of Psychological Research Online, 8*(2), 23-74.

Sipahi, B., Yurtkoru, E., & Çinko, M. (2010). *Sosyal bilimlerde SPSS'le Veri Analizi.* İstanbul: Beta Basım A.Ş.

Tabachnick, B., & Fidell, L. (2001). *Using multivariate statistics.* Boston: Ally and Bacon.

Taşkın, Ç., & Akat, Ö. (2010). *Araştırma Yöntemlerinde Yapısal Eşitlik Modelleme.* Bursa: Ekin Basım Yayın.

Widaman. (1985). Hierarchically Tested Covariance Structure Models for Multitrait-Multimethod Data. *Applied Psychological Measurement, 9*, 1-26.

Wooldridge, J. (2003). *Introductory Econometrics: A Modern Approach* . Mason: Thomson.

Wu, A., & Zumbo , B. (2008). Understanding and Using Mediators and Moderators. *Social Indicators Research, 87*, 367-392.

Wu, J.-H. W., & Wang, Y.-M. (2006). Measuring KMS Success: A Respecification of the DeLone and McLean's Model. *Information & Management, 43*, 728-739.

Yung, Y., & Bentler, P. (1994). Bootstrap-corrected ADF test statistics in covariance structure analysis. *British Journal of Mathematical and Statistical Psychology, 47*, 63-84.